The New Fire Safety Legislation 2006 – A Practical Guide

Pete Muir

Acknowledgements

The author and publishers wish to thank the following for permission to reproduce copyright material:

The Chief Fire Officer's Association (CFOA) for kind permission to reproduce example notices and associated documentation

DCLG and Crown publishers for kind permission to reproduce text from the RR(FS)O

Crown copyright material is reproduced with permission of the controller of HMSO and the Queen's Printer for Scotland

Published by the Royal Institution of Chartered Surveyors
under the RICS Books imprint
Surveyor Court
Westwood Business Park
Coventry CV4 8JE
UK
www.ricsbooks.com

ISBN 978-1-84219-309-9

Typeset in Great Britain by Columns Design Ltd, Reading, Berkshire.
Printed in Great Britain by Cromwell Press, Trowbridge, Wiltshire

Contents

Preface

Over four years in the making, the *Regulatory Reform (Fire Safety) Order* 2005 ('the RR(FS)O') is set to change the face of fire safety in England and Wales from October 2006.

Fire safety is to become forward thinking, proactive and pre-emptive, as opposed to the traditionally reactive approach to historical fire events.

Out go the *Fire Precautions (Workplace) Regulations* 1997 along with the 40-year-old Fire Certification Scheme. Over 100 other pieces of existing legislation are affected in some way or another by the RR(FS)O.

The RR(FS)O has been designed to be the single, overriding piece of fire safety legislation applicable to premises of almost every kind, with only a few very specific exceptions.

No longer is the employer the only person to be held responsible for fire safety of premises. The new scope of premises to which the order applies means that a person who has control over any part of the premises is now burdened with the onus of being a 'responsible' person for the fire safety of the premises over which they have control.

There are many publications already in existence which guide the reader through the steps to carry out fire risk assessments and optimise the fire safety of their premises. The Department for Communities and Local Government (DCLG) have published a series of 11 guides aimed at 'responsible' persons in specific premises and which provide invaluable information relating to the provision of fire safety and the carrying out of fire risk assessments as required by the RR(FS)O.

This book presents the reader with an overview of the new fire safety legislation, how it is to be implemented and what effect it is likely to have. It is hoped that the reader will be left in no doubt as to their role in relation to fire safety, how the new legislation will be enforced and how it may be complied with.

It presents the reader with an introduction to the processes the enforcing authority may use to determine inspection priority, etc. and also provides examples of the types of enforcement actions which may be taken as a result of any legislative infringement. By removing the sense of 'mysticism' surrounding the role of the enforcing authority, it is hoped that the reader will be less apprehensive in approaching the relevant bodies for advice prior to the Local Fire and Rescue Services (LFRS) carrying out any inspection of their premises. There is a wealth of knowledge and experience available to anyone who requires it and the LFRS has a duty to supply fire safety advice and guidance when requested.

Bringing together the people who are responsible for fire safety in premises and those who have a knowledge of fire safety can result in massive benefits for all concerned – fewer premises will pose unacceptable fire risks to occupants (i.e. fewer fire incident casualties) and the job of the LFRS becomes so much easier.

This book specifically targets those people who may be in doubt as to whether they are directly affected by the introduction of the RR(FS)O. Property managers at all levels, employers, landlords and events organisers, to name but a few, will gain something from reading this book.

If this book in any way serves to promote fire safety awareness or if it helps to break down any of the perceived barriers between the public and the authorities then its publication is justified and the effort in producing it proven worthwhile.

Acknowledgements

To Sabrina Kuhn of RICS Books for continually hounding me to get the manuscript across for editorial review. I am sure that publication of this book would not have happened if it were not for her dedication to the cause.

To Cheryl Prophett of Proof Positive for kind words of encouragement and the patience and ability to sift through the author's drivel and create something readable in the English language.

To Maria Finegan, Managing Director of Safe Consulting Ltd for allowing me the time to pen the manuscript for this book.

To Alex Soltynski and all my other colleagues at SAFE Consulting Ltd for their contributions and continual support of this project.

Finally – to my wife Sue and our two girls, Lisa and Sarah. How they endured my 'hijacking' space and time in penning this text I do not know; without their support this book could not have been completed.

Pete Muir (September 2006)

Introduction

Made on 7 June 2005, the *Regulatory Reform (Fire Safety) Order*, hereafter referred to simply as the RR(FS)O, comes into force from 1 October 2006.

It is the intention of this book to provide an insight into the reasoning behind the RR(FS)O and, perhaps more importantly, the major implications its implementation demands.

Previous UK fire safety legislation has been somewhat disjointed and to a certain degree drafted and subsequently introduced as a 'knee jerk' reaction to fire events where there has been major loss of life. From the Great fire of London in 1666 (actually this was the second great fire of London – the first occurring in 1212 with loss of life estimated at anything between 3,000 and 12,000 depending on which sources you read!) where six people lost their lives but many thousands more were made homeless, the early monarchy, and subsequently governments, have been scripting and passing Acts and Regulations intended to provide ever higher levels of protection for the general populace from the risks of fire.

A 1993 Home Office review identified that the then current fire safety legislation did not provide a suitable means for progressing fire safety into the future; subsequently, several years of review, discussion and debate between various 'interested parties' has spawned what we now know as the RR(FS)O.

In the meantime, whilst the reviews and consultations were taking place, existing legislation (specifically the *Fire Precautions (Workplace) Regulations* 1997) was amended to place a direct responsibility for the fire safety of employees on the shoulders of the employer. The concept of risk assessment was introduced as a way of demonstrating that due cognisance had been taken by an employer of the possible risk of fire and its effects on his/her workforce.

The Fire Regulations (together with the *Health and Safety at Work etc. Act* 1974) required an employer to carry out a fire risk assessment of the workplace and, if there were more than five employees normally on or in the premises, keep a written record of any such assessments and actions required/carried out as a result.

The RR(FS)O continues along the same lines and anyone already involved in workplace fire risk assessments will be familiar with the process involved to achieve compliance with the new legislation.

The RR(FS)O has the effect of repealing part of the *Fire Precautions Act* 1971 and revoking the *Fire Precautions (Workplace) Regulations* 1997, as well as making amendments to numerous other pieces of legislation. As a result, the existing Fire Certification scheme is being abolished and previously granted certificates will cease to have any legal status; however, any information contained within the fire certificate document package will prove useful as a starting point for the fire risk assessments required under the RR(FS)O.

One other important implication of the RR(FS)O is the definition of premises and those to which the order applies. There is a huge increase in the diversity of premises to which a fire risk assessment must be applied; in fact, rather than providing a list of premises to which the order applies, it provides a concise list of premises to which the RR(FS)O and its requirements need not be applied.

In short, the RR(FS)O is intended to be the single piece of legislation which covers all aspects of fire safety in so far as the vast majority of premises are concerned – the following being notable exceptions:

- single private dwelling domestic premises (limited provision is made for other domestic premises);
- offshore installations;
- ships, but only with regard to crew operations;
- agricultural or forestry fields, wood or land not inside a building and situated away from the main buildings;
- a number of means of transportation;
- a mine development (above-ground buildings are not exempt from the order);
- borehole sites.

Further discussions relating to the definition of the term 'premises' in relation to the RR(FS)O and the application thereof to various buildings, etc. take place in Chapter 2.

Whilst this book does have a chapter dedicated to fire risk assessment, it is not the author's intention to provide a comprehensive guide to fire hazards and methods of mitigation for all situations. The Department for Communities and Local Government (DCLG) has produced a series of 11 guides specifically aimed at different types of premises and it is recommended that reference be made to such documentation for more detailed guidance on the type of fire risk assessment required, and how to effectively carry it out. The premises covered by the guides are as follows:

- offices and shops;
- premises providing sleeping accommodation;
- residential care;
- small and medium places of assembly;
- large places of assembly;
- factories and warehouses;
- theatres and cinemas;
- educational premises;
- healthcare premises;
- transport premises and facilities;
- open air events.

The guides provide information on general fire safety for each of the property types and, if followed by a competent person, should provide the user with sufficient information to conduct a fire risk assessment of sufficient detail to ensure that all possible necessary steps have been taken in regard to fire safety.

This book is divided to provide distinct sections of information and guidance related to specific aspects of the RR(FS)O implementation:

- fire legislation history;
- the RR(FS)O;
- roles and responsibilities;
- fire risk assessments;
- fire engineering;
- enforcement;
- offences and appeals;
- miscellaneous provisions;
- UK legislation.

Where excerpts from UK fire safety legislation have been reproduced within the text of this book, they are provided as supplementary information and are 'boxed' out, from the main body of text. The book is written in such a way as to permit the reader to 'skip' these inserts but still gain an understanding of what is required of him/her in relation to the provisions required by the RR(FS)O. Also included at the back of this book is the text from the RR(FS)O (except Schedules 1–5), as it appears on the DCLG website.

All new buildings must comply with the current Building Regulations and the DCLG has published a series of Approved Documents in support of the regulations, which present building designers, etc. with a set of recommendations which if followed will lead to compliance with the regulations. Of particular interest to the reader of this book is Approved Document B (ADB) – fire safety. It should be noted at this point that certain older buildings may not comply with the current Building Regulations; however, any subsequent 'material change' to the building (be that by use, occupation or structure) will require the building to be upgraded to meet the current regulations. In the case of a renovated building where no 'material change' has taken place, Building Regulation 4 is applicable which demands that the result of renovations/changes do not make the fire safety provisions of the building any worse for the intended occupants than were evident prior to the renovations/changes.

This book contains a chapter entitled simply 'Fire Engineering'. This may seem a little out of context when discussing fire safety legislation, however the UK Building Regulations have for some time permitted (and encouraged) fire safety provisions which fall outside of the scope of ADB in order to maximise flexibility of building design and use. Not everyone is conversant with the principles of fire engineering; without this knowledge, a building may be looked upon as non-compliant with the Building Regulations simply because the fire safety measures provided do not follow the recommendations contained within ADB. This chapter of the book seeks to provide a basic insight into fire engineering, the associated guidance available, and demonstrate a few typical examples which may be encountered in 'the real world'. In no way does this book purport to explain the intricacies involved in the production of fire engineered solutions, but it should give the reader an insight into a small number of alternatives to the approved document approach to fire safety. If there is any doubt as to whether a building complies with the Building Regulations where it does not necessarily follow the advice of ADB then specialist assistance should be sought from either one of the Local Fire and Rescue Service Fire Engineering Groups (current contact details included in Appendix D), or a dedicated fire engineering consultant.

Of the issues surrounding the implementation of the RR(FS)O, the definition and assignment of the 'responsible person' is crucial. It is the duty of such a person to ensure the fire safety of the premises for which they are responsible; this places the onus of accountability firmly on the shoulders of those whose position encompasses the role. The guidance contained within this book will aid in the determination of such responsibilities and to whom they apply, depending on the circumstances of each situation. In placing the responsibility for fire safety on a person, there is the inevitable need to 'police' the provisions required by the RR(FS)O. The chapter entitled 'Enforcement' outlines the role of Building Control and Local Fire and Rescue Services in this regard.

1 A brief history of major UK fire legislation

It would be unfair to start this book without firstly examining a brief history of Fire Safety Legislation in the UK. It is hoped that the reader (perhaps despairing over the recent changes) will gain an understanding of the driving factors behind changes in the law and also perhaps encourage an empathy with those involved in the decision to make this latest change a reality.

THE BEGINNINGS

The origins of much of the existing (2005) Fire Safety Legislation can be traced back to an event in which fire played a major part some time before the publication of the legislation. The general attitude appears to have been one of the future prevention of prior events, rather like the old adage of 'hindsight engineering', or 'stable door' engineering. To this end we have been graced with a wealth of legislation which addresses, or attempts to address, Fire Safety in some respect or other. We now find ourselves in the unenviable position of having over 100 pieces of legislation which contain Fire Safety instruction/information/guidance.

Let's take a look at where some of the multitude of Fire Safety Legislation actually came from.

1189

The first evidence of Fire Legislation in the UK dates from around 1189 when the then Mayor of London declared that all houses should be constructed from stone and effectively banned thatched roofs. This drastically predates the commonly assumed beginnings of Fire Safety Legislation by almost 500 years.

1212–1666

The Great Fire of London in 1666 is frequently cited as the reason that Fire Safety Legislation was introduced in the UK. Not only was the 1666 event not the original instigator of Fire Safety Legislation in the UK it was also the second Great Fire of London. The first recorded 'Great Fire' of London actually occurred in 1212 and, depending on which source you read, the number of lives lost ranges in estimate from 3,000 to 12,000 and anywhere in between. The 1666 fire accounted for the official loss of only six lives (and the displacement of many thousands) but due to the year (news was an issue and events were starting to be documented more than ever before) it is still regarded as 'The Great Fire of London'.

It is true to say, however, that the 1666 fire has proven to be the catalyst for an upsurge in Fire Safety awareness; beginning with King Charles II proclamation that the walls of all new buildings should be constructed of stone, narrow streets should be widened to lessen the threat of fire spread between opposing buildings and a further decree that the narrow alleyways of London should be reduced in number.

1774

The first real fire legislation came into force somewhat later as *The Fire Prevention (Metropolis) Act* of 1774. It took another half a century and a number of serious fires before the formation of the Municipal Fire Brigades of Edinburgh and Manchester in 1824 and 1828 respectively; followed shortly thereafter (1840s onwards) by the introduction of further fire safety related legislation.

1844–1858

The *Metropolitan Building Act* reduced the number of building classes from seven (in the *Metropolis Act*) to three. The subsequent *Towns Improvement Clauses Act* set out standard clauses for local improvement Acts, e.g. construction of roofs and walls, etc.

1850

This year saw the introduction of the *Burgh Police Act* which stated that party walls had to be carried through the roof line and, further, that external roofs and walls were to be constructed from non-combustible material.

1858

At last, local authorities had the power to create legislation which was tailored to their own demographic and geographic needs.

1861–1865

The 1861 Tooley Street warehouse fire in London prompts the publication of the *Metropolitan Fire Brigade Act* of 1865.

1871–1877

Following consultation, the Local Government Board commences drafting *Model Building Byelaws* which were to be published in 1877.

1890

Introduction of the *Public Health Act Amendments Act* which provides guidance on minimum levels of safety for places of assembly with particular regard to means of escape.

1891

The *Factories and Workshop Act* describes control of means of escape requirements for factories with in excess of 40 employees. It also introduces protected measures for means of escape from tall buildings (over 60ft high) and adds a requirement for Local Council Certification of the building prior to occupation.

1895

The *Factories and Workshops Act* is amended to include factories where there are less than 40 employees.

1905

The *London Building Acts (Amendment) Act* required work to be carried out on certain existing premises to facilitate escape in the event of a fire. Plans now required to be submitted for all new building work and a schedule of 'fire-resisting materials' is published.

1921

A Royal Commission is appointed to review the existing Fire Safety Legislation/provisions.

1930

The *London Building Acts* 1930 introduce some 39 separate powers to enable the generation of byelaws.

1934

The Home Office introduces the long-awaited 'Manual of Safety Requirements in Theatres and other places of Public Entertainment'.

1937

The *Factories Act* introduced a requirement to provide suitable means of escape from affected industrial buildings and enforcement was cited as being the responsibility of the local Council Sanitary Inspector.

1938

Fire Brigades Act.

1947

The *Fire Services Act* saw the first piece of legislation which placed a burden of responsibility on the Fire Brigade for disseminating, on request, advice in respect of aspects of fire safety; prevention, restriction of fire spread and means of escape.

1948

The *National Assistance Act* contained a single paragraph relating to the provision of fire and accident safety measures to be applied to nursing and old people's homes. The level of provision was to be determined according to the mental and physical condition of the occupants of such premises.

1959

Following the Eastwood Mills, Keighley fire in 1956 where eight people lost their lives and the subsequent Public Enquiry the *Factories Act* was amended to incorporate additional and improved safety provisions to form the 1959 Act.

1961

The consolidation of the means of escape provision from the 1937 Act and the safety provisions of the 1959 Act into the 1961 *Factories Act* meant that the Act now encompassed fire alarms, fire-fighting provisions, fire resistant construction, and outwardly opening doors. The local fire authority became responsible for the provision of a means of escape certificate for the premises whereas the Factory Inspector was responsible for other provisions required by the Act.

1963

The *Office, Shops and Railway Premises Act* was penned primarily as a result of the 1960 Henderson's department store fire in Liverpool where 11 people lost their lives. Unfortunately (at least as far as the Fire Brigade were concerned), although requirements were made for various fire safety provisions in each of the affected premises, only certain of the premises would actually require a Fire Certificate, depending upon size and occupancy. Inevitably this led initially to a substantial number of 'small premises' not being subject to Local Fire Brigade inspection as they would not be required to hold a valid Fire Certificate.

1970

Holroyd Committee separates Legislation for building versus occupation.

1971

The *Fire Precautions (Places of Work) Act* was undoubtedly one of the most important Acts ever placed on the statute books in so far as the issue of fire safety is concerned. Here we had the first Act of Parliament which was targeted solely at the provision of fire safety and fire prevention measures.

The Act was written in such a way as to enable its application to a wide range of premises, not being restricted only to those requiring a Fire Certificate.

1972

The *Fire Precautions (Hotels and Boarding House) Act* was the first Act with a direct and tangible link to the

Fire Precautions Act 1971 and made it mandatory for an establishment with sleeping accommodation for more than six (including staff) to hold a valid Fire Certificate. There were other types of premises requiring Fire Certification such as those with sleeping accommodation at below ground level (i.e. basements), or with sleeping accommodation for guests or staff at above first floor level. This new designation was largely a reaction to the 1969 Crown Hotel fire in which 11 people died.

1973

For the first time, the Building Regulations contained a section devoted to Structural Fire Precautions which also provided requirements for the means of escape in case of fire.

1980

The *Upholstered Furniture Regulations* were introduced as a direct response to the Woolworths store fire in Manchester which claimed the lives of ten people, mainly from smoke inhalation.

1987

The *Fire Safety and Safety at Places of Sport Act* was brought about as a direct result of the 1985 Bradford City Football Ground tragedy in which some 56 people lost their lives. The Act defined fire safety objectives for the affected premises, in addition to decreeing certain sports premises as requiring a valid Fire Certificate.

1989

The *Fire Precautions (Sub Surface Railway Stations) Regulations Act* was introduced in an attempt to prevent a recurrence of the 1987 King's Cross disaster which saw the loss of some 31 lives.

The *Fire Precautions (Factories, Offices, Shops and Railway Premises) Order* came into effect in April, replacing the 1976 Order. The prime purpose of the Order was to remove the exemption for self-employed persons and the term 'employed to work' was replaced with 'at work'.

This action and consequent reactionary process has continued right into the mid-1990s where, eventually, it was decided that a full review of UK Fire Safety Legislation was required.

1993

The Home Office review of the *Fire Precautions Act* 1971 concluded that the current Act did not provide the most suitable legislative means of ensuring fire safety at the time or in the future.

1994

An interdepartmental Review of Fire Safety Legislation and Enforcement examines fire safety arrangements across government, and recommends a modernisation and rationalisation of the legislative and organisational framework. The Review recommended that general fire safety in the workplace (covered by the *Fire Precautions Act* 1971) should fall under the same legislative regime as process fire safety, i.e. under the *Health and Safety at Work etc. Act* 1974. The proposals were rejected and separate legislation was continued.

1997

Consultation paper, *Fire Safety Legislation for the Future*, is issued which described a 'radical overhaul' of existing legislation, and introduced a 'new, modern approach based on risk assessment'.

Fire Precautions (Workplace) Regulations introduced as a result of EEC directive.

1999

Amendment to *Fire Precautions (Workplace) Regulations* and the introduction of workplace fire risk assessments. This provided the first major change in direction for fire safety provisions, requiring workplaces to have a person or persons who could be held accountable for the fire safety of the workplace occupants.

2000

As a result of the 1993 review a number of review and audit teams and commissions were set up with the eventual formation of the Fire Safety Advisory Board which was to provide an open forum for fire safety discussions the aim of which was to provide a series of recommendations for change.

The *Regulatory Reform Act* was enacted in 2000 and was intended to permit the reform of legislation. This Act was followed within two years by the Consultation Document on the Reform of Fire Safety Legislation.

2002

Consultation document on the proposed reform of fire safety legislation published

A consultation document containing draft regulatory changes was distributed to 'interested parties' and comments collated and, where relevant, incorporated prior to bringing the proposed new legislation before the two Houses of Parliament.

2004

On 10 May 2004 the government laid before Parliament a draft proposal for the *Regulatory Reform (Fire Safety) Order* 2004 in the form of a draft of an Order and an explanatory memorandum from the then Office of the Deputy Prime Minister (renamed as the Department for Communities and Local Government mid-2006).

The proposal is intended to achieve a wide-ranging consolidation of the existing legislation relating to fire safety, thereby reducing the overlap of existing multiple fire safety regimes and the overlap of enforcing authorities. The aim of the proposed order is to achieve a single regulatory regime for fire safety, with one authority in each area responsible for enforcement of general fire safety issues.

THE NEED FOR CHANGE

Virtually all previous fire legislation and guidance has been based on a prescriptive approach to the issue of fire safety and whilst, for the applications it was intended for, such legislation has proved its worth there existed a certain rigidity which prevented its application to more complex situations. One further drawback to most existing guidance and legislation (certainly prior to the introduction of risk assessments) is the inability to embrace technological advances in fire science, fire detection and fire protection methods.

With the recent introduction of risk assessment to fire safety issues, the wheels were set in motion for a change in direction for fire safety policy and planning. No longer should fire safety be driven by past events; fire safety was set to become pre-emptive and proactive, rather than just reactive.

The introduction of a risk assessment based approach to fire safety brings with it a freedom never before experienced in this field; strict adherence to the approved documents is no longer a requirement, thereby opening the doors for greater flexibility in design and fire safety provisions.

Risk assessments provide the tools to identify hazards and permit us to provide mitigation against such risks. The old adage of 'prevention is better than cure' has never been more appropriate.

Instead of different legislation dealing with different situations, we now have a single document which applies to virtually all premises, and requires the application of the same process to each.

In a nutshell:

- Previous legislation has been reactive – generally in response to a historical major fire incident.
- The RR(FS)O provides a proactive, risk-based approach which provides a suitable vehicle for moving fire safety into the future.

2 The RR(FS)O

In this chapter we will take a look at just what the RR(FS)O is, its implications, and application to premises.

The RR(FS)O is the result of a number of years' work in reviewing existing legislation, drafting proposals and consulting with interested parties, and has endured a number of amendments and readings before both Houses of Parliament before finally being drafted for implementation.

The RR(FS)O comprises five parts, containing 53 Articles and five further Schedules, the five parts are:

Part 1 General
Part 2 Fire safety duties
Part 3 Enforcement
Part 4 Offences and appeals
Part 5 Miscellaneous

And the five schedules are:

Schedule 1 Risk assessment, prevention and dangerous substances (broken into four parts)
Schedule 2 Amendments of primary legislation
Schedule 3 Amendments of sub-ordinate legislation
Schedule 4 Repeals
Schedule 5 Revocations

By necessity, this book is concerned only with the contents of Parts 1 to 5 and the 53 Articles contained therein. The additional five Schedules are 'tacked' on to the main legislative document and provide informative guidance and reference to other affected legislation. The purpose of this book is primarily to clarify the reader's position in relation to the provision of fire safety measures in his or her premises and as such the five schedules, whilst informative, are not discussed herein; the reader is directed to the full text of the RR(FS)O which may be freely downloaded from the DCLG website and numerous sources on the Internet, or purchased from The Stationery Office (TSO).

Whilst the changes made to the *Fire Precautions (Workplace) Regulations* 1999 went largely unnoticed by the general public (and to an extent largely ignored), this is certainly not the case with the implementation of the RR(FS)O. The government and Local Fire and Rescue Services have embarked on an extensive publicity campaign, including the leafleting of all potentially affected premises.

The main effect of the changes is a shift in direction towards greater emphasis on fire prevention in all non-domestic premises, including the voluntary sector and self-employed people with premises separate from their homes.

The fire certification scheme will be scrapped and existing certificates will cease to have any legal status.

The RR(FS)O applies only to premises in England and Wales. (Northern Ireland and Scotland will have their own, similar legislation.) It covers 'general fire precautions' and other fire safety duties which are needed to protect 'relevant persons' in case of fire in and around most 'premises'. The RR(FS)O requires fire precautions to be put in place 'where necessary' and to the extent that it is reasonable and practicable in the circumstances of each case.

The 'responsible person' will be required to carry out a fire risk assessment which must focus on the safety of all 'relevant persons' in the event of a fire. It should pay particular attention to those at special risk, such as the disabled and those with special needs, and include assessment of any hazardous or dangerous substance likely to be on the premises. The fire risk assessment will identify risks that can be removed or reduced and assist in the decision as to what general fire precautions require implementation in order to protect people against the fire risks that remain. If there are more than five people employed on the premises any significant findings of the assessment must be recorded. Refer to Chapter 4 for some general guidance relating to fire risk assessments.

The RR(FS)O does not add anything particularly new to fire safety law, instead it consolidates and replaces existing legislation to cover a greater number of premises types.

Responsibility for compliance with the RR(FS)O will rest with a 'responsible person'. In a workplace, this is generally the employer and any other person who may have control of any part of the premises, e.g. the occupier or owner. In all other premises the person or people in control of the premises will be responsible. If there is more than one responsible person in any type of premises, each must take all reasonable steps to work with each other. Refer to Chapter 3 for further information relating to the 'responsible person' and their implied duties under the RR(FS)O.

The RR(FS)O makes mention of general fire precautions for the premises and outlines a series of headings/topics into which the subject may be split:

> ***Duty to take general fire precautions***
> ***8.*** *—(1) The responsible person must—*
> *(a) take such general fire precautions as will ensure, so far as is reasonably practicable, the safety of any of his employees; and*
> *(b) in relation to relevant persons who are not his employees, take such general fire precautions as may reasonably be required in the circumstances of the case to ensure that the premises are safe.*

GENERAL FIRE PRECAUTIONS

The RR(FS)O details what is meant by 'general fire precautions':

> ***Meaning of "general fire precautions"***
> ***4.*** *—(1) In this Order "general fire precautions" in relation to premises means, subject to paragraph (2)—*

Reducing the risk of fire and spread of fire

> *(a) measures to reduce the risk of fire on the premises and the risk of the spread of fire on the premises;*

This is a very broad requirement which places an onus to limit the amount of combustible material on the premises to a minimum practical level and encompasses the provision of fire rated compartments, structures, etc.

Reduction of the risk of fire encompasses a number of possible measures which may be employed, some of which are listed below:

- reduction or removal of flammable and combustible inventory;
- reduction or removal of sources of ignition;
- separation of fuel and potential sources of ignition;
- implementation of active controls where working with flammable or combustible materials is unavoidable;
- implementation of early warning automatic fire detection systems; this measure will not reduce the probability of a fire occurring but will permit early movement of occupants and intervention in the event of a fire thus reducing any risk from the fire.

Fire spread can occur in a number of ways.

- Convection – Heat is carried in currents which give rise to the familiar 'fire plume' with its characteristic 'V' shape. Smoke and other products of combustion carry heat with them as they leave the fire and this heat can then be transferred to the building's internal structure and materials.
- Conduction – The direct transfer of heat through a material; for example, a fire occurring on one side of a metal door could soon raise the temperature of the other side of the door to a point where it may cause ignition of other materials in the vicinity.
- Radiation – The major reason for fire spread. Radiation is the prime mover behind a fire being self-sustaining or accelerating. Heat from flames is radiated back into the fuel, which in turn heats up to the point where it pyrolyses and produces more flammable vapour to take part in the combustion process. Radiation is also regarded as the predominant cause of flashover, the point at which a fire accelerates rapidly to engulf and involve an entire room or compartment; a hot layer of smoke and gas in the upper part of the room radiates heat back to the floor and contents of the room, eventually (but not always) elevating the temperature of any combustible material to the point where it ignites.

Reducing the risk of fire spread therefore involves the application of numerous possible measures depending on the perceived risk.

- Compartmentation – the provision of construction elements which are designed to withstand the effects of fire for a predetermined period of time.
- Separation of fuel sources to the extent that a fire in one part of a building or room will not cause other combustible materials to ignite.
- Separation of buildings by either fire-resisting construction or by a suitable distance to prevent the spread of fire from one to the other.
- The introduction of active fire control measures such as sprinkler systems to limit the size of any fire occurring on the premises.
- Providing hot smoke extraction to control the upper hot smoke layer temperature such that the probability of a flashover event is substantially reduced.

It is not within the scope of this book to explain in any detail the mechanics surrounding heat transmission, pyrolysis, combustion, fire growth, etc.; there are many books which deal with these subjects in great detail. If the reader desires a basic understanding of the principles of fire, a suitable starting point would be the Local Fire and Rescue Services, many of whom run excellent courses aimed at the non-scientific interested party.

Back to the RR(FS)O: the requirement does not detail the measures which may be employed, it is the duty of the 'responsible person' to demonstrate that all practical steps have been taken, and measures implemented to limit the risk and spread of fire on the premises. Thus is it important that the 'responsible person' (or their delegated representative) has at least a basic understanding of how fires start, spread and how they may be controlled.

Means of escape

(b) measures in relation to the means of escape from the premises;

There must be sufficient means of escape routes, of sufficient capacity, to cater for the postulated occupancy of the premises.

In many cases, especially newly built premises, the capacity of routes out of the building may be subject to the minimum requirements for disabled access; this may necessitate the provision of route widths which are greater than would necessarily be required for fire escape purposes, although the reader is reminded that this may not always be the case and many existing buildings may exhibit seemingly unacceptably narrow escape/exit provisions.

(c) measures for securing that, at all material times, the means of escape can be safely and effectively used;

Means of escape routes, exits and stairs must be free from obstructions and blockages such that they remain available for use at all material times.

Fire doors and other doors on escape routes should be easily openable from the side approached in the event of a required evacuation and not require special tools (e.g. a key) or specialist training in their use. There are exceptions to this, depending on occupancy type, numbers and management procedures.

This requirement also relates to the provision of adequate emergency escape signage and emergency lighting (BS5499 and BS EN50172 provide guidance), after all, if occupants cannot see the route or exit they cannot use it!

Provisions for fire fighting

(d) measures in relation to the means for fighting fires on the premises;

Relating to the provision of hand-held fire-fighting equipment such as hoses, portable extinguishers, fire blankets and the like.

It must be ensured that the correct extinguisher type is provided for the perceived fire risk in the area, and that sufficient are provided to enable fighting of small fires. Staff must be made aware of the dangers of using the wrong extinguisher type on certain fires (e.g. water extinguishers and electrical fires are contra-indicated) and there should always be someone present in the premises who has received training specific to the operation of hand-held fire extinguishers.

The requirement in certain buildings for the provision of fire-fighting stairs and lifts, wet or dry fire mains, etc. falls under this section. The 'responsible person' is required to ensure that the installations are maintained in such a condition that they can be called upon for use in the event of a fire; periodic inspection and testing is essential in this regard.

This section may also be applied to the use of fixed fire suppression systems such as sprinklers, for example, especially where they may be required by legislative guidelines; certain Local Acts require the installation of sprinkler systems in buildings which may be over a certain height, or which may have an enlarged footprint area.

Fire detection and warning

(e) measures in relation to the means for detecting fire on the premises and giving warning in case of fire on the premises; and

A suitable method of raising an alarm must be maintained for the premises. This may not necessarily require the provision of an Automatic Fire Detection (AFD) system; in small premises it may be sufficient from a life safety viewpoint to have a manually operated horn/sounder by an exit door. For single room premises it may be acceptable for the first person to see the fire simply to shout the alarm.

The more complex the premises are, and the higher the perceived risk to occupants is (never mind potential material/capital losses), the greater the need will be for some form of AFD.

Recommendations on AFD type, provision and level of coverage are contained in numerous reference documents and standards (BS5588 series for example) and guidance on the design and installation of AFD systems can be found in BS5839, Part 1.

It is imperative that occupants of a building are briefed on whatever system of fire detection and alarm is employed, how to recognise the alarm, manually initiate an alarm, and what to do upon hearing an alarm.

Correctly specified, designed and installed AFD and alarm systems can provide warning of a fire usually in the incipient stages, often before the first flame has been generated. This can provide the time required for manual intervention and fire-fighting prior to any major damage occurring.

Actions to be taken in the event of a fire

> *(f) measures in relation to the arrangements for action to be taken in the event of fire on the premises, including—*
> *(i) measures relating to the instruction and training of employees; and*
> *(ii) measures to mitigate the effects of the fire.*

Occupants of the premises must be informed of the fire safety measures in place and also what action to take in the event of a fire alarm being raised. This includes the provision of suitable fire safety plan notices throughout the premises. Occupants are expected to be made aware of their escape routes and assembly points and have fire drills carried out at regular intervals.

It is one of the duties bestowed on the 'responsible person' to ensure that all occupants receive an appropriate level of fire safety training pertaining to their premises. Records should be kept of all relevant training undertaken.

Training takes many forms and varies in extent depending on the particular circumstances of the premises and occupants. The 'responsible person' for a small shop, for example, may find it satisfactory to carry out training of staff themselves, and this may take the form of verbal instruction on the measures to be taken in the event of a fire.

Conversely, the 'responsible person' for a larger supermarket may require delegates from their staff to attend organised fire safety awareness/management courses, etc. Instruction particular to the premises concerned and the actions to be taken in the event of a fire would still form an essential part of any training regime implemented.

In all cases the level of training, and the selection of occupants to receive the training, will need to be assessed by the 'responsible person' for the premises; advice regarding the type and level of training required may be obtained from the Local Fire and Rescue Services; a selection of contact details are listed towards the end of this book.

In a nutshell:

- The RR(FS)O requires the 'responsible person' to take appropriate steps to ensure the fire safety of people on the affected premises.
- The RR(FS)O does not detail what is meant by 'reasonably practicable', the onus is on the 'responsible person' to show that they could not have reasonably been expected to do/provide more for the fire safety of the occupants.
- The Local Fire and Rescue Service are there to help you and can provide advice on all matters related to fire safety.

EFFECTS ON OTHER LEGISLATION

The RR(FS)O is intended to replace all other existing fire-related legislation and has the effect of repealing part of the *Fire Precautions Act* 1971 and revoking the *Fire Precautions (Workplace) Regulations* 1997, as well as making amendments to numerous (estimated at over 100) other pieces of legislation.

To illustrate the extent to which the RR(FS)O affects other legislation the following is a non-exhaustive list of affected documents:

Amendments

Celluloid and Cinematograph Film Act 1922
London Building Acts (Amendment) Act 1939
East Ham Corporation Act 1957
Caravan Sites and Control of Development Act 1960
Public Health Act 1961
Gaming Act 1968
Fire Precautions Act 1971
Health and Safety at Work etc. Act 1974
Safety of Sports Grounds Act 1975
Greater London Council (General Powers) Act 1975
County of South Glamorgan Act 1976
Rent Act 1977
County of Merseyside Act 1980
West Midlands County Council Act 1980
Cheshire County Council Act 1980
West Yorkshire Act 1980
Isle of Wight Act 1980
South Yorkshire Act 1980
Tyne and Wear Act 1980
Zoo Licensing Act 1981
Greater Manchester Act 1981
County of Kent Act 1981
Derbyshire Act 1981
East Sussex Act 1981
Local Government (Miscellaneous Provisions) Act 1982
Humberside Act 1982
County of Avon Act 1982
Cumbria Act 1982
Hampshire Act 1983
Staffordshire Act 1983
Food Act 1984
Building Act 1984
County of Lancashire Act 1984
Cornwall County Council Act 1984
Bournemouth Borough Council Act 1985
Leicestershire Act 1985
Worcester City Council Act 1985
Poole Borough Council Act 1986
Berkshire Act 1986
Fire Safety and Safety of Places of Sport Act 1987
Plymouth City Council Act 1987
West Glamorgan Act 1987
Dyfed Act 1987

Environment and Safety Information Act 1988
Smoke Detectors Act 1991
London Local Authorities Act 1995
Capital Allowances Act 2001
Licensing Act 2003
The Marriages (Approved Premises) Regulations 1995
The Construction (Health, Safety and Welfare) Regulations 1996
The Housing (Fire Safety in Houses in Multiple Occupation) Order 1997
The Health and Safety (Enforcing Authority) Regulations 1998
The Building Regulations 2000
The Building (Approved Inspectors etc.) Regulations 2000
The Care Homes Regulations 2001
The Children's Homes Regulations 2001
The Private and Voluntary Care (England) Regulations 2001
The Care Homes (Wales) Regulations 2002
The Private and Voluntary Care (Wales) Regulations 2002
The Children's Homes (Wales) Regulations 2002
The Child Minding and Day Care (Wales) Regulations 2002
The Residential Family Centres Regulations 2002
The Residential Family Centres (Wales) Regulations 2003

It should be noted that in many cases, local acts confer particular requirements for fire safety in premises located in the area covered by the Act; the RR(FS)O *does not* remove these requirements but should be seen as the one overriding piece of legislation to which the various local acts provide supplement. It is important that the 'responsible person' at least make themselves aware of the existence of any local acts and the implications they may bring. Once more, advice on any aspect of fire safety may be obtained from the Local Fire and Rescue Services.

Repeals

London Building Acts (Amendment) Act 1939, c.xcvii
East Ham Corporation Act 1957, c. xxxvii
Gaming Act 1968, c. 65
Fire Precautions Act 1971, c. 40
Fire Precautions (Loans) Act 1973, c. 11
Health and Safety at Work etc. Act 1974, c. 37
Safety of Sports Grounds Act 1975, c. 52
Greater London Council (General Powers) Act 1975, c. xxx
County of South Glamorgan Act 1976, c. xxxv
Rent Act 1977, c. 42
Local Government, Planning and Land Act 1980, c. 65
County of Merseyside Act 1980, c. x
West Midlands County Council Act 1980, c. xi
Cheshire County Council Act 1980, c. xiii
West Yorkshire Act 1980, c. xiv
Isle of Wight Act 1980, c. xv
South Yorkshire Act 1980 c. xxxvii
Greater Manchester Act 1981, c. ix
County of Kent Act 1981, c. xviii
Derbyshire Act 1981, c. xxxiv
East Sussex Act 1981, c. xxv
Local Government (Miscellaneous Provisions) Act 1982, c. 30
Humberside Act 1982, c. iii
Cumbria Act 1982, c. xv
Hampshire Act 1983, c. v
Staffordshire Act 1983, c. xviii
Food Act 1984, c. 30
Building Act 1984, c. 55
County of Lancashire Act 1984, c. xxi
Bournemouth Borough Act 1985, c. v
Leicestershire Act 1985, c. xvii
Clwyd County Council Act 1985, c. xliv
Worcester City Council Act 1985, c. lxiii
Poole Borough Council Act 1986, c. i
Berkshire Act 1986, c. ii
Fire Safety and Safety of Places of Sport Act 1987, c. 27
Plymouth City Council Act 1987, c. iv
Dyfed Act 1987, c. xxiv
Environment and Safety Information Act 1988, c. 30
National Health Service and Community Care Act 1990, c. 19
Smoke Detectors Act 1991, c. 37
Capital Allowances Act 2001, c. 2

Revocations

The Fire Certificate (Special Premises) Regulations 1976
The Fire Precautions (Workplace) Regulations 1997
The Fire Precautions (Workplace) (Amendment) Regulations 1999
The Management of Health and Safety at Work Regulations 1999

In the context of this book, the terms 'repeal' and 'revocation' have identical meaning; i.e. the legislation to which the terms are applied will no longer have legal status once the RR(FS)O is implemented.

In a nutshell:

- The intention is to provide a single piece of legislation which can be applied unilaterally to all premises with the exception of single occupancy domestic dwellings.
- Local acts will still provide a supplement to main fire safety legislation.

FIRE CERTIFICATES

Pre-October 2006 position

The use of certain types of premises had been designated as requiring a fire certificate under the *Fire Precautions Act* 1971.

There have been two pieces of legislation which required the provision of Fire Certificates in Great Britain – one relates to larger hotels and boarding houses; the other to those factories, offices, shops and railway premises in which people are employed to work.

The first piece of legislation (the *Fire Precautions (Hotels and Boarding Houses) Act* 1972) requires a fire certificate when premises are used as a hotel or boarding house which will provide sleeping accommodation for more than six people (whether employees or guests) or if they provide sleeping accommodation for employees or guests elsewhere than on the ground or first floors of the premises.

The second piece (the *Fire Precautions (Factories, Offices, Shops and Railway Premises) Order* 1989) requires that a fire certificate be applied for when more than 20 people are at work at any one time in the workplace, or more than ten are at work at any one time elsewhere than on the ground floor.

In many cases where the requirements of the Fire Regulations had been complied with, this would have been sufficient for the fire and rescue authority to have issued a fire certificate without any further action.

Post-October 2006

The RR(FS)O scraps the 40-year old fire certification scheme. The scheme is not being replaced by anything new; it is now the duty of the 'responsible person' for the premises to ensure that occupants are as safe from the affects of a fire as is practical. This does not imply a lesser responsibility for the safety of occupancy of the premises; it is almost certain that for premises which required a fire certificate under the old scheme, similar fire safety measures will be required under the RR(FS)O.

From January 2006 a transitional period was defined and some Local Authorities ceased issuing new fire certificates for premises from this time. Just to confirm the views of some about the lack of consistency in the interpretation of current fire legislation and guidance, some authorities have decided to issue new fire certificates right up to the point where the RR(FS)O comes into force. The debate for consistent approach across the country could fill a library (and then some), therefore this book will concentrate on the effects of the RR(FS)O on businesses, occupants and others, rather than the perceived right and wrong way of its interpretation and implementation

Existing fire certificates and supporting files and documentation should not simply be discarded when the RR(FS)O is implemented; the information contained therein provides an excellent basis from which a fire risk assessment can be carried out. It is, in fact, likely that premises with existing fire certificates will already have a fire risk assessment in place and only minor amendments will be required to comply with the RR(FS)O.

In a nutshell:

- Just because the fire certification scheme is being scrapped, do not think there is no longer a responsibility for ensuring the fire safety of people on the premises.
- Don't throw away the old fire certificate documentation; use it as a foundation for future risk assessment.
- The Local Fire and Rescue Service are there to help you and can provide advice on all matters related to fire safety.

WILL MY BUILDING BE SUBJECT TO THE RR(FS)O?

In short, the most probable answer to this question is yes, so please read on.

The RR(FS)O applies to far more premises than any previous fire safety legislation. This is intentional; it is not only workers and employees who are at risk from fire within a building, any risk present may be equally applicable to all occupants.

This section of the book also looks at the premises to which the RR(FS)O applies, those to which it does not, and how the Fire Service may apply risk assessment to prioritise their inspection plans.

Definition of premises

The RR(FS)O is very specific in this regard and rather than attempt to define premises to which the Order applies, simply states which particular premises are not subject to application of the RR(FS)O.

Application to premises

6. *—(1) This Order does not apply in relation to —*

(a) domestic premises, except to the extent mentioned in article 31(10);

(b an offshore installation within the meaning of regulation 3 of the Offshore Installation and Pipeline Works (Management and Administration) Regulations 1995;

(c) a ship, in respect of the normal ship-board activities of a ship's crew which are carried out solely by the crew under the direction of the master;

> *(d) fields, woods or other land forming part of an agricultural or forestry undertaking but which is not inside a building and is situated away from the undertaking's main buildings;*
> *(e) an aircraft, locomotive or rolling stock, trailer or semi-trailer used as a means of transport or a vehicle for which a licence is in force under the Vehicle Excise and Registration Act 1994 or a vehicle exempted from duty under that Act;*
> *(f) a mine within the meaning of section 180 of the Mines and Quarries Act 1954, other than any building on the surface at a mine;*
> *(g) a borehole site to which the Borehole Sites and Operations Regulations 1995 apply.*

The inclusion of domestic premises is purposeful, as is the definition thereof:

> *"domestic premises" means premises occupied as a private dwelling (including any garden, yard, garage, outhouse, or other appurtenance of such premises which is not used in common by the occupants of more than one such dwelling);*

Further, article 31(10) states:

> *In this article, "premises" includes domestic premises other than premises consisting of or comprised in a house which is occupied as a single private dwelling and article 27 (powers of inspectors) shall be construed accordingly.*

Both statements effectively limit the definition of 'domestic premises' and by inference bring premises such as HMOs, care homes, and the like, under the influence of the RR(FS)O.

The RR(FS)O defines premises as follows:

> *"premises" includes any place and, in particular, includes—*
> *(a) any workplace;*
> *(b) any vehicle, vessel, aircraft or hovercraft;*
> *(c) any installation on land (including the foreshore and other land intermittently covered by water), and any other installation (whether floating, or resting on the seabed or the subsoil thereof, or resting on other land covered with water or the subsoil thereof); and*
> *(d) any tent or movable structure;*

Tent? – *TENT*? – Yes, really. Tents are specifically mentioned along with movable structures such as site cabins, etc. because they are no longer solely used by campers or ramblers. Tents, by definition, include any structure or erection, comprising a support structure and weatherproof membrane. The structure may be steel, aluminium, carbon fibre or any other material and may even be compressed air (as in the case of self-supporting inflatable tents). The weatherproof membrane may be simply 'draped' over the supporting structure, slung underneath, or even form part of the structure.

What really matters is the use to which the tent is put. More and more, quickly erected tents are used throughout the country as temporary (sometimes less temporary than you may think – Millennium Dome?) accommodation for all manner of activities. In all respects, the safety of those occupying these spaces must be protected as if they were in a conventional building.

So, going back to the defined premises we notice the inclusion of vehicles, vessels, aircraft and hovercraft. Premises? Well, yes, in so far as they are all modes of transport capable of functioning as a workplace or place of public occupation – particularly when stationary. The inclusion of these now means that a boat, moored or dry docked and used as, say, a restaurant or museum will be subject to application of the RR(FS)O. The use of old rolling stock (and aircraft to a lesser degree) for purposes other than transport has increased over recent times (particularly within the licensed trade) and are now included.

Ships are treated in a slightly different manner; only the normal activities of the ship's crew under direction from the master are exempted from application of the RR(FS)O. All other activities (e.g. if the ship is a cruise ship or ferry with members of the public on board) will be subject to the Order.

Fields, woods or other land forming part of an agricultural or forestry undertaking – this seems quite sensible and extremely obvious, but there is a specific purpose for this single, open air exemption; that is to render all other open air 'premises' (for example an open air music venue) subject to the provisions of the Order.

RR(FS)O Applicability

The following table lists the more commonly encountered types of premises, together with an indication as to whether the RR(FS)O is applicable to each:

Place	Subject to RR(FS)O	Comment
Airport	YES	
Amusement arcade	YES	
Animal slaughterhouse	YES	
Art gallery	YES	
Bank or building society	YES	
Bingo hall	YES	
Boarding house/guesthouse	YES	
Bus/coach station/terminus	YES	
Call centre	YES	
Car park	YES	
Car showroom	YES	
Caravan site	YES	
Care home	YES	
Cargo vessel	NO	Subject to IMO and other regulations
Casino	YES	
Catering marquee	YES	
Cinema	YES	
Common areas within blocks of flats, etc.	YES	
Concert hall	YES	
Conference centre	YES	
Cruise ship	YES	
Educational establishment	YES	
Exhibition centre	YES	
Factory	YES	
Farm	YES	
Fire station	YES	
Fishing vessel	NO	Subject to IMO and other regulations
Flat or maisonette (single dwelling)	NO	
Funfair	YES	
Halls of residence	YES	
Health centres and surgeries	YES	
HMO	YES	
Hospital	YES	
Hostel	YES	
Hotel/motel	YES	
House (single dwelling)	NO	

Place	Subject to RR(FS)O	Comment
Laboratory	YES	
Law court	YES	
Leisure complex	YES	
Library	YES	
Meeting hall	YES	
Museum	YES	
Non-nuclear power generation installation	YES	
Nuclear installation	YES	Also subject to UKAEA and other regulations
Nursery and day care centre	YES	
Office	YES	
Performing arts venue	YES	
Place of detention (e.g. prisons, detention centres and certain police stations)	YES	But note dissapplication of s.14(2)(f) for security reasons
Place of worship	YES	
Police station	YES	
Post office	YES	
Restaurant	YES	
Retail premises	YES	
Riding school	YES	
Sea-going port	YES	
Skating rink	YES	
Sports venue	YES	
Swimming pool	YES	
Temporary (for example site cabin type) accommodation	YES	
Theatre	YES	
Tourist attraction (any)	YES	
Train station/terminus	YES	
TV, radio or music studio	YES	
Warehouse	YES	
Workshop	YES	
Zoo	YES	

The comments relating to places of detention should be relatively self-explanatory – security of the occupants has to take precedence. However, the safety of the occupants from the effects of fire must still be maintained, thus the remainder of the RR(FS)O applies.

So – returning once more to the original question *'Will my building be subject to the RR(FS)O?'* – if you are not talking about the dwelling in which you live (and in some circumstances even if you are!), and you cannot find it in the short list of exemptions contained within the RR(FS)O, the answer is almost certainly *YES*.

You will effectively assume responsibility for the fire safety of the occupants of the building, in those areas which are to any extent under your control.

In a nutshell:

- Yes – the RR(FS)O applies to your premises, unless they fall into the short list of exceptions.
- If any areas of the premises are to any degree under your control, you will assume the status of 'responsible person'.
- Single occupancy dwellings are excluded.

3 Roles and responsibilities

WHOSE JOB IS IT ANYWAY?

The RR(FS)O places a burden of responsibility firmly on the head of a 'responsible person' with regard to the fire safety of the occupants of the premises to which they have been assigned.

The responsible person is required to co-ordinate all fire safety related issues including the carrying out of a fire risk assessment and production of associated documentation.

The responsible person may, if required by the RR(FS)O, nominate 'competent persons' to assist in the implementation of any measures deemed necessary to ensure the fire safety of the occupants of the premises. The responsible person must also ensure that any such competent persons have received an adequate level of training to allow proper implementation of any measures related to their fire safety related activities. Although the onus of responsibility for the assessment, provision and maintenance of fire safety measures may be placed on a third party, the legal responsibility for the fire safety of the property still lies with the 'responsible person' as defined in article 3 of the RR(FS)O.

THE RESPONSIBLE PERSON

The RR(FS)O appears quite clear as to who the responsible person should be:

> *(a) in relation to a workplace, the employer, if the workplace is to any extent under his control;*
> *(b) in relation to any premises not falling within paragraph (a)—*
> *(i) the person who has control of the premises (as occupier or otherwise) in connection with the carrying out by him of a trade, business or undertaking (for profit or not); or*
> *(ii) the owner, where the person in control of the premises does not have control in connection with the carrying on by that person of a trade, business or other undertaking.*

This may be relatively straightforward in the case of, for example, a small privately owned shop where the proprietor has complete control over the fire safety provisions within the premises. Clearly, the proprietor is the responsible person and must ensure the safety from fire of his/her employees and patrons of the shop.

The matter of deciding who the 'responsible person' may be becomes somewhat complex where there are a number of organisations or businesses occupying the premises.

For example, in a multi-occupancy office block, each organisation or business operating from that building would require a responsible person for their part of the building (irrespective of whether that part of the building is owned by or leased to the occupant) if they have any control over the fire safety provisions within the premises. The Landlord/Owner would by default be the responsible person for all Landlord controlled/common areas of the building.

Where a contract agreement between a Landlord and lessee includes for the Landlord to take care of maintenance and/or repair of the premises, or any part thereof, the Landlord is deemed to have control of the premises in so far as the extent of his control of the premises. What this means is that the Landlord will essentially not be responsible for the operations/actions of the individual tenants and their approach toward fire safety, but would be responsible for the provision and maintenance, for example, of sufficient fire exits, etc. Here we have a shared responsibility, the landlord is responsible for the material provisions relating to fire safety, whereas the tenant is responsible for the fire safety detail specific to his/her operations within the premises.

A further complication to the identification of the 'responsible person' can occur when an employer takes space in a building in which the fire safety measures are not in his control. The owner may also not have control over the fire safety measures, at which point the onus of responsibility passes back to the employer to resolve any fire safety issues with the Landlord, building owner or their agent. In short, it is likely that any employer will have ultimate responsibility for the fire safety of their employees and any other persons who may occasion the premises from which they operate.

There will be a number of issues raised when the RR(FS)O comes into force. For example: a Landlord has a building which he has fitted with a category M fire alarm system. His prospective tenant carries out a risk assessment and decides that a minimum system requirement would be L2. Who is responsible for the implementation of the system? The tenant because he deems it necessary or the Landlord because the existing system is under his control? According to the above, if the tenant is an employer, the responsibility for fire safety passes back to him/her for resolution.

Taking a step backwards, let's assume that the premises in question are yet to be built. Is there a responsibility to ensure that the fire safety provisions are adequate?

The short answer is yes. The RR(FS)O does not alter the fact that buildings must be designed and built to meet the functional requirements of the Building Regulations. The RR(FS)O requires fire risk assessments for buildings predominantly when they are, or are about to be, occupied. Unless the design of the building follows the prescriptive recommendations contained within ADB, there will be eventual difficulties in performing a fire risk assessment if the details of any non-ADB solutions are not known. It is imperative therefore, that the designer/owner/planning applicant compiles a fire safety strategy for the building which details all fire safety measures and how/why they are implemented (Chapter 5 provides an insight into some non-ADB solutions and fire engineering). Not only will such a document be extremely useful during the planning stage, it provides information vital to the understanding of how the building 'works' and will assist the undertaking of a comprehensive fire risk assessment.

RELEVANT PERSONS

For the purposes of the RR(FS)O, a 'relevant person' is any person who may be put at risk by, or affected by, the occurrence of fire within the premises. The definition not only applies to those on or in the premises but also covers persons who may be in the immediate vicinity of the premises and may be adversely affected by a fire in or on the premises.

There is a responsibility to provide as safe an environment as possible for all who may frequent the premises. Also, do not forget that the 'responsible person' described earlier is also a 'relevant person'.

In a nutshell:

- 'Responsible person' – anyone who has a measure of control over the premises; no longer just the employer.
- 'Relevant person' – anyone who may occasion the premises whether as an employee or otherwise. Also applies to persons in the vicinity of the premises who may be affected by a premises fire incident.

4 Fire risk assessment

WHY FIRE RISK ASSESSMENT?

The requirement for fire risk assessments to be carried out has been retained from previous legislation and, in essence, remains unchanged. Risk assessment for certain premises has been around for the last seven years or so but never has the subject been more publicised than with the implementation of the RR(FS)O.

Although the risk assessment approach is retained, the scope of premises to which the requirement applies has been broadened dramatically. The RR(FS)O does not distinguish between premises type or location in so far as fire risk assessment is concerned; the new system now requires ALL premises covered by the RR(FS)O (refer back to Chapter 2) to have had carried out a fire risk assessment by the 'responsible person' or any other nominated 'competent person'.

The requirement for making and keeping records of any fire risk assessment remains, where the premises contain five or more employees, where a licence is in force in relation to the operation of the premises or any activities therein, or where a notice made under the RR(FS)O requires it.

With the shift of emphasis within the RR(FS)O to a risk based approach to life safety of all occupants it is the author's recommendation that even if there are fewer than five employees on the premises, but five or more persons during normal business operations (a small supermarket or a dental or doctor's surgery for example) that a documented fire risk assessment should be made:

> ***Risk assessment***
>
> *9. —(1) The responsible person must make a suitable and sufficient assessment of the risks to which relevant persons are exposed for the purpose of identifying the general fire precautions he needs to take to comply with the requirements and prohibitions imposed on him by or under this Order.*

The decision as to whether a recorded fire risk assessment is required or not rests ultimately with the Local Fire and Rescue Service, whose job it is to determine the adequacy of proof that the premises (for their intended use) comply with the requirements of the RR(FS)O. In many cases a recorded fire risk assessment may be the *only* method of proving that the requirements of article 8(1)(a), (b) and 9 have been met.

Many premises, certainly those operating under a fire certificate, will already have had a fire risk assessment carried out under previous legislation and therefore the RR(FS)O will have little impact upon them. The RR(FS)O now extends the fire risk assessment (not necessarily recorded) requirement to virtually all premises where people are, or may be, present, irrespective of numbers or activities.

Fire risk assessments should be carried out either by the responsible person, or by another nominated 'competent person'. In any case the person carrying out the assessment must be able to demonstrate their competence to carry out the task by one or more of the following means:

- qualification – having studied, sat and attained a recognised qualification in an appropriate subject;
- experience – have a proven (and traceable) long term history in the field of fire safety;
- incorporation – be a member of a recognised body in the field of fire safety (for example the register of fire risk assessors and auditors from the Institution of Fire Engineers);
- by any other means recognised as acceptable by the relevant Local Fire and Rescue Service.

If in any doubt at all, employers and other 'responsible persons' are advised to contact their Local Fire and Rescue Service for initial advice before embarking on the process of carrying out a fire risk assessment of their premises.

The level of detail required for a fire risk assessment to be deemed adequate for the premises to which it applies is dependent on a number of factors:

- the number of persons at risk from any fire;
- the nature of the business or activities involving the premises;
- the materials/processes contained on the premises;
- the demeanour of the occupants of the premises and their ability or willingness to react in an emergency;
- the effect a fire on the premises would have on the immediate environment.

To illustrate the different levels of detail in fire risk assessments for different premises, a selection of examples are provided in Appendix B.

FIVE STEPS TO RISK ASSESSMENT

The DCLG has for some time advocated the implementation of the 'five steps to risk assessment' based on original HSE guidance (INDG 163). This methodology has stood the test of time and is worth revisiting here.

Step 1: Identify all potential fire hazards
Step 2: Identify all persons at risk
Step 3: Evaluate the risk and apply the five Rs
Step 4: Record all findings and actions
Step 5: Review the risk assessment regularly

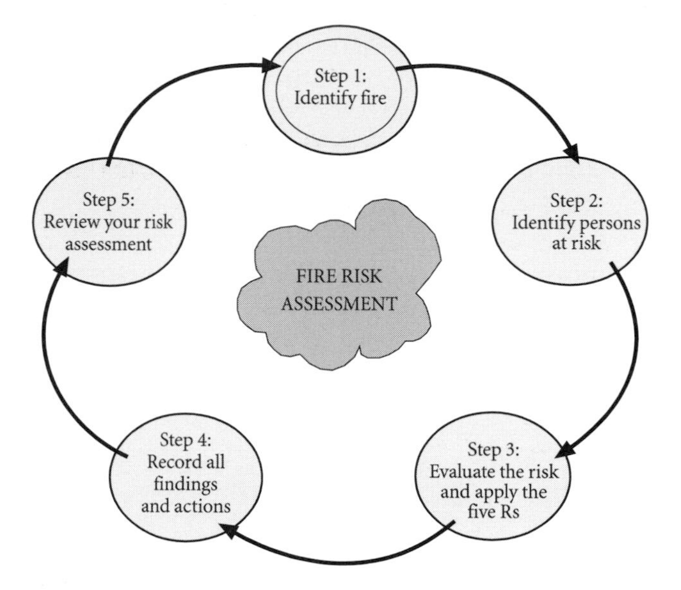

As can be seen in the diagram above, fire risk assessment is not a single, one shot process; rather it is a lifetime process which requires review on a regular basis (recommended annually, or sooner should the use or nature of the premises change).

At the risk of repeating guidance from many other excellent documents available, each of the five steps are briefly outlined below.

Fire hazards

Fundamental to the understanding of fire risks is an appreciation of the chemical reaction we call fire.

Fire is an exothermic chemical reaction which occurs between a fuel and an oxidising agent (usually the oxygen in air) which results in the release of large amounts of energy, generally in the form of heat and light. The reaction is rarely spontaneous and normally requires initiation (ignition) by an external source.

As with everything in life however, there are exceptions to the rules.

- A hydrogen fire produces heat but no visible flame.
- Discarded rags soaked in oil based paint for example can 'burst into flames' with no external ignition source.
- Coal (stock piled) is known to smoulder under specific conditions, again with no apparent external ignition source.
- Haystacks, affected by moisture and biological reactions, can have their core temperatures rise to the point at which combustion occurs.

Most people reading this book will have heard of the 'Fire Triangle' in one guise or another:

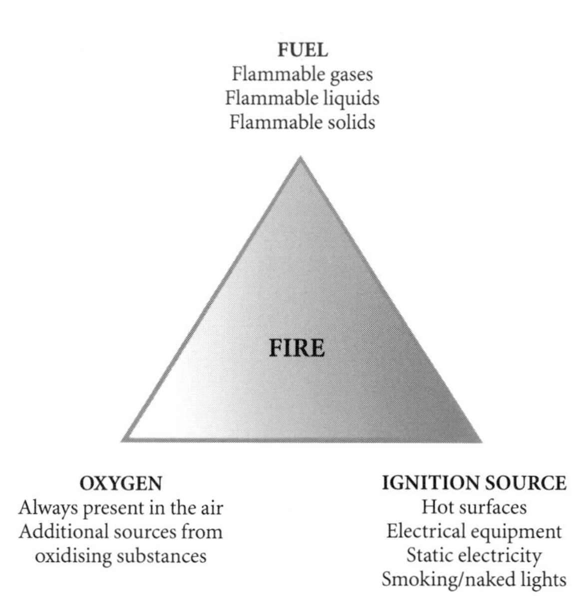

Whilst the fire triangle does little to explain the actual mechanics of fire, it provides sufficient understanding of the principle elements required in order for a fire to occur. Each of the three elements of the fire triangle represents a fire risk category which must be dealt with as part of any fire risk assessment.

- Fuel
  - Perhaps an obvious hazard but consideration should be given to all flammable and combustible materials. Look for:
    - waste paper, wood, sawdust, etc.;
    - stored combustible materials, plastics, wood, paper, card, etc.;
    - display materials;
    - furniture coverings;
    - paints, oils, varnishes and other solvents or solvent based products;
    - flammable gases such as propane and butane (commercially known as LPG);
    - foam-filled products, plastics and polystyrene, etc.
- Ignition source
  - Again a seemingly obvious hazard, consideration should be given to all of the following:

- smoking materials, cigarettes, lighters, matches, etc.;
- any naked flames (including candles, blow torches, etc.);
- faulty, damaged, or misused electrical equipment and cabling;
- proximity of lights to possible fuels;
- cooking facilities;
- hot processes such as welding, torching, etc.;
- obstructed vents on electrical and other equipment;
- heaters of any kind.

- Oxygen
  - Not such an obvious hazard but there are many processes which use an oxygen enriched atmosphere, medical treatments often include the administration of oxygen, and there are also numerous chemicals which are classified as oxidising agents.

The lists above do not attempt to identify all the possible fire risks but should provide an insight into the more common categories of hazards.

Persons at risk

All persons who may occasion the premises must be protected from the affects of fire, not only employees. The risk assessment must identify numbers and locations of all potential premises' occupants.

Particular attention should be paid to those who may be particularly vulnerable and who fall under any of the following categories:

- the young;
- the old or infirm;
- those with disabilities (not only movement impaired, but also those with sight and/or hearing impediments);
- those who may, as part of their day-to-day operations, be in proximity to a potential fire hazard;
- people who may be unfamiliar with the premises, e.g. visitors, customers, contractors, etc.;
- people who may be isolated or work alone in the premises, cleaners, security personnel, etc.

The above occupants (or potential occupants) will, in most cases, require special measures to be implemented to assure them of an adequate level of safety from fire. Such measures may range from simple instruction relating to fire alarms and evacuation, to the introduction of physical means of ensuring their safety.

It is important that each category of occupant be assessed separately as fire safety needs will differ from group to group depending on their individual needs and ability to react in the event of a required evacuation.

Evaluating the risk

The next stage in the fire risk assessment is to evaluate the risks identified, both individually and collectively. This exercise initially requires a 'two pronged' approach:

- evaluation of the risk of a fire occurring; and
- evaluation of the risk to people.

The combination of risk of occurrence and risk to people varies from situation to situation and from premises to premises; for example, it may be perfectly acceptable for an able bodied person, trained in the dangers and hazards of a process using flammable materials, to be exposed to the hazard during their working day. Conversely, it would not be acceptable for a young person to be present, unaccompanied, in, say, a plant room where access and therefore means of escape was restricted. Each combined risk will require assessment in its own right to determine whether the level of risk posed to the occupant is acceptable or not.

When assessing the overall risk, consideration should be given to the following fire safety provisions which, depending on the level of provision, may have an effect on the level of the perceived risk:

- means of escape;
- building fire compartmentation;
- fire detection and alarm;
- fire suppression systems;
- portable fire extinguishing facilities;
- training of personnel in fire safety;
- emergency escape signage;
- emergency lighting;
- access for the Fire Service.

The above list is not exhaustive, but each and every one of the measures listed will affect the level of risk from a building or process to the individual concerned.

Any fire risks identified should have the five Rs applied to it; this represents a hierarchical approach to the provision of mitigation against the risk and should be applied in 'top-down' order.

The five Rs, in order, are:

- Remove
  - Can the fire hazard be removed from the premises completely?
  - Can ignition sources and combustibles be kept apart?
  - Ensure that a no smoking policy is strictly enforced.
  - Can waste/refuse storage be kept apart from the premises?
  - Can occupants be separated from fire risk areas?
- Replace
  - Can the substance/process posing the fire hazard be replaced with one which presents a lower or no risk?
  - Can heaters, etc. with naked flames be replaced with 'safer' alternatives?

- Reduce
  - Can the substance/process posing the risk be reduced – i.e. less flammable inventory?
  - Can occupants be limited in their exposure to fire risks?
  - Provide AFD system to provide early warning of fire.
- Regulate
  - Control the risk by the use of hazard safety measures.
  - Provide automatic fire suppression.
  - Ensure that any combustible stock, etc. is strictly controlled
- Restrict
  - Restrict use or access to the risk by means of administrative processes (e.g. permit to work system).

The five Rs are represented in the diagram below:

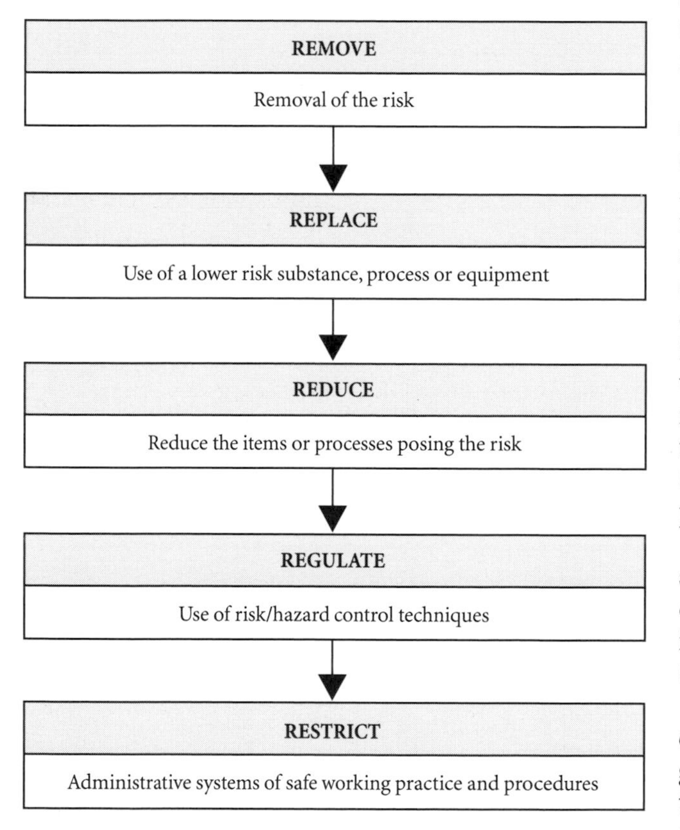

Record keeping

In our present climate of QA enlightenment and the 'where there's blame there's a claim' culture it is important that the results of any risk assessment be recorded for future reference; any actions resulting from the assessment should also be recorded.

Although for certain premises and occupancies legislation would not require the fire risk assessment to be recorded, it is good practice to do so and may provide useful reference should the use or occupancy of the premises change in the future.

In the event of the premises being subject to an audit by the Fire Service it is important to be able to demonstrate that fire risks have been assessed and reduced to as low a level as practical under the given circumstances.

Review

This aspect cannot be stressed enough – risk assessment is not a 'one-shot' exercise; it is a cyclic event which feeds back into itself.

It is recommended that fire risk assessments take place at intervals of one year, or whenever there has been a perceived change in the fire risk to the building occupants, whichever is sooner. It is not necessary to re-evaluate the risk assessment each time a minor change occurs (say the addition of a server room to an office building), but major changes to the building, its occupants or the material use of the building should initiate a review of the fire safety measures in place.

The level of detail to which a fire risk assessment should go is beyond the scope of this book (a small selection of assessments are provided in Appendix B), and indeed the RR(FS)O. Suffice to say that the choice of fire risk assessment should be appropriate for the risk perceived; the outlined assessment above may be suitable for most retail, office, commercial and light industrial situations, but it would not be suitable, or acceptable for situations where the processing and handling of flammable materials takes place on a day-to-day basis, or where the potential for consequential loss is extremely high (e.g. power generation installations, petrochemical plants, etc.).

Fire risk assessment methodology has been the subject of exhaustive works and reference material can be obtained from any number of sources including the DCLG, Health and Safety Executive (HSE) and the Local Fire Service.

The DCLG, together with substantial input from external interested parties, has produced a series of guides aimed at the 'responsible person' for different types of premises. The guides promote a logical process of risk assessment and mitigation, together with useful information relating to fire safety principles in general.

In a nutshell:

- Risk assessment is not a 'one shot' exercise – regular reviews are necessary.
- Identify fire hazards.
- Identify persons at risk.
- Evaluate and mitigate.
- Record findings and actions.
- Review, review, review.

FURTHER READING

The following documents, relating to fire risk assessment, are published by the DCLG and may be purchased from reputable booksellers, or downloaded from the Internet by visiting www.communities.gov.uk and following the links provided.

A short guide to making your premises safe from fire
Product code 05 FRSD 03546

Fire Safety – Risk Assessment
Office and shops
ISBN 1851128150

Fire Safety – Risk Assessment
Factories and warehouses
ISBN 1851128167

Fire Safety – Risk Assessment
Sleeping accommodation
ISBN 1851128174

Fire Safety – Risk Assessment
Residential care premises
ISBN 1851128181

Fire Safety – Risk Assessment
Educational premises
ISBN 1851128198

Fire Safety – Risk Assessment
Small and medium places of assembly
ISBN 1851128204

Fire Safety – Risk Assessment
Large places of assembly
ISBN 1851128211

Fire Safety – Risk Assessment
Theatres and cinemas
ISBN 1851128228

Fire Safety – Risk Assessment
Outdoor events
ISBN 1851128239

Fire Safety – Risk Assessment
Healthcare premises
ISBN 1851128242

Fire Safety – Risk Assessment
Transport premises and facilities
ISBN 1851128255

5 Fire engineering

WHAT IS FIRE ENGINEERING?

Most people, when asked what fire engineering means, will give one of three replies:

(1) 'I don't know.'
(2) 'What a fireman does.'
(3) 'Sprinklers and fire alarms.'

In fact, fire engineering is recognised as a science in its own right and there are a number of universities around the country offering degree courses on the subject. Fire engineering covers a range of topics including, but not limited to, the following:

- principles and behaviour (science) of fire;
- how fires start;
- reaction of materials to fire;
- reaction of people to fire;
- production and behaviour of products of combustion (including smoke);
- fire spread within and between buildings;
- fire control by compartmentation;
- fire control by active protection measures;
- fire detection;
- fire extinguishment;
- interaction of fire safety and other building systems;
- application of 'outside of the box' solutions to Building Regulations compliance issues (i.e. not relying on the recommendations of ADB and other prescriptive guidance).

WHY USE FIRE ENGINEERING?

In general terms, fire engineering can be described as the determination and application of fire safety principles which meet the functional requirements of the Building Regulations, but do not necessarily comply with prescriptive recommendations.

Indeed ADB states in its first few pages:

> *The Approved Documents are intended to provide guidance for some of the more common building situations. However, there may well be alternative ways of achieving compliance with the requirements.*
>
> ***Thus there is no obligation to adopt any particular solution contained in an Approved Document if you prefer to meet the relevant requirement in some other way.***

and ...

> ***Fire safety engineering***
>
> ***0.11*** *Fire safety engineering can provide an alternative approach to fire safety. It may be the only practical way to achieve a satisfactory standard of fire safety in some large and complex buildings, and in buildings containing different uses, e.g. airport terminals. Fire safety engineering may also be suitable for solving a problem with an aspect of the building design which otherwise follows the provisions in this document.*

As the Building Regulations provide performance related criteria in respect of fire safety, the recommendations (and that is, after all, what they are – recommendations) contained within ADB are not the only method by which compliance can be achieved.

Thus there is a plethora of solutions which may be applied to any given situation requiring compliance with the Building Regulations, where application of the recommendations in ADB may not be desirable or even possible.

It is important that building occupiers who may take on the role of 'responsible person' are aware of any fire engineered principles which may have been applied to the fire safety provisions of the premises.

It is not within the scope of this book to discuss the principles of fire engineering in great depth. There are numerous existing texts dealing with the subject (some conflicting and some controversial) and there are a number of fire engineering companies upon whose wealth of knowledge the reader could draw.

EXAMPLES

The following page illustrates a few examples where the fire engineering principles applied to the premises may not be immediately obvious to the 'responsible person'.

When a property is occupied, there *must* be a duty of care on behalf of the owner or landlord to provide the 'responsible person' for the premises with information relevant to and detailing any deviations from the recommendations contained within prescriptive guidance.

The application of correct fire engineering principles can result in a building which has a fire safety level equal to or better than the prescriptive recommendations, but which may not be evident without prior knowledge of the basis upon which the fire safety measures were designed.

The duty placed on the 'responsible person' should not be underestimated; they need to be fully aware of current fire legislation, the fire safety provisions for the premises in which they are involved, and any fire engineered solutions which may have been applied to the premises. In this respect a 'fire strategy' document for the building, which describes in detail any fire engineered principles applied, is crucial for the 'responsible person' when performing a fire risk assessment. This document will provide sound arguments and reasoning for the fire safety of the building including how and why it complies, not necessarily with ADB, but with the Building Regulations.

If in any doubt at all, the 'responsible person' should seek advice from their Local Fire and Rescue Service, the building designer or from a reputable fire engineering company.

In a nutshell:

- Fire engineering provides an alternative to the ADB approach to Building Regulations compliance.
- Fire engineering may not be obvious to the casual observer.
- Fire safety strategy documentation is essential to the understanding of any fire-engineered approach to fire safety for the premises.

Examples where fire engineering has resulted in a different approach to fire safety than the prescriptive guidance contained in ADB

- Reduction of required fire protection from 120 minutes to main structural elements
- Removal of requirement for fire protection to secondary structural members
- Use of 'over-size' smoke reservoirs in high roof building
- Use of extended travel distances
- Use of service corridors as part of means of escape and fire fighting access
- Reduction of the number of escape stairs compared to prescriptive recommendations
- Reduction or removal of fire rating requirement for atria glazing
- Reduction or removal of fire rating requirement for building facade glazing
- Reduction of the number of fire-fighting shafts compared to prescriptive recommendations
- Use of timber and similar materials for internal facades to atria
- Use of 'progressive horizontal evacuation' traditionally reserved for healthcare premises
- Reduction or removal of prescriptive requirement for sprinkler fire protection system
- Increased compartment sizes when compared to prescriptive recommendations

Note:

Photos of buildings are for illustration purposes only and do not infer that any of the above fire engineered solutions have been applied specifically to any of the premises pictured

6 Enforcement

It is necessary, when any piece of legislation is to be applied successfully, that a responsibility for enforcement be clearly defined; nowhere is this more important than when the application is intended to secure the safety of life.

The RR(FS)O places the onus of enforcement into five discreet categories, dependent upon the type of premises and the use to which the premises are, or are intended, to be put.

WHO ARE THE ENFORCERS?

It should be noted at this point that a Fire and Rescue Service, being appointed as the enforcing authority, may make arrangements with either the Health and Safety Commission or the Office of Rail Regulation, as appropriate, to perform some or all of the duties of the enforcing authority.

An inspector working directly for, or on behalf of, the enforcing authority is bestowed with the power to:

- enter premises, without force, as necessary for the performing of his duties;
- enquire and ascertain the compliance of the premises with the RR(FS)O, including the identification of the responsible person;
- demand the production of documentation which would be required as a result of compliance with the order, or which he deems necessary in order to perform his duties under the order and to take copies thereof;
- demand facilitation and assistance in the carrying out of his duties from the responsible person in so far as their responsibilities extend;
- remove samples of any articles/materials from the premises for the purpose of assessing their fire resistance or flammability;
- require the dismantling or testing of anything found on the premises which may, in his opinion, have been the cause or is likely to be a cause of danger to relevant persons.

The inspector must:

- produce, on demand, proof of his identity and authority in relation to the RR(FS)O;
- if requested by a person who has responsibilities for the premises, only carry out actions arising from the order in the presence of said person;
- prior to requiring the dismantling or testing of any article or substance believed to present a danger to relevant persons, seek consultation with appropriate persons to ascertain what dangers may be posed by carrying out such an action.

A fire inspector, or other person authorised by the Secretary of State may, with the consent of the Local Fire and Rescue Services, authorise a member of the Fire and Rescue Services to assume the role of inspector (and in doing so, take on the same powers under the RR(FS)O) in relation to an affected premises or any part thereof.

The chart on the next page provides a simplified approach to deciding who the enforcing authority is likely to be for most types of premises.

BUILDING CONTROL AND PLANNING APPLICATIONS

Enforcement of the RR(FS)O does not simply rely upon the relevant authority carrying out inspections of premises and issuing notices where required. For any new building the planning application approvals process is where the requirements of the order will be first encountered.

One of the requirements of the RR(FS)O is a responsibility placed upon the local authority (in most cases the Local Council Building Control Department) to enter into consultation with the enforcing authority (in the majority of cases the Local Fire and Rescue Services, assuming that the local authority is not also the enforcing authority) prior to consenting to approval of any plans deposited with them.

During any planning application there will normally be three main consultations between the Building Control Officer and the Fire and Rescue Service which may be categorised as follows:

- initial, informal advice;
- detailed consultation for plans approval;
- pre-completion consultation.

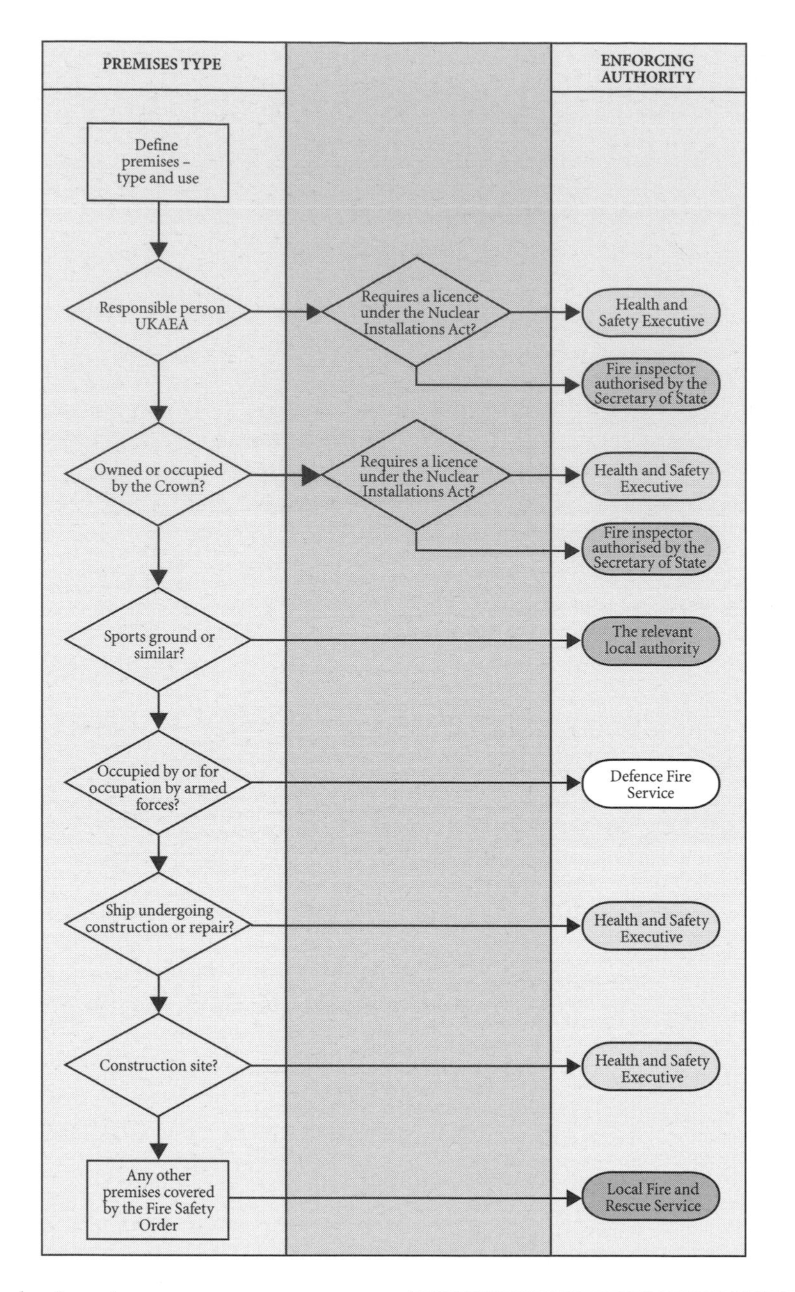

Each round of consultations is necessary to ensure a smooth passage through the approvals process and compliance with both the functional requirements of Building Regulations and the RR(FS)O.

Initial advice may however be sought by the designer from the Building Control Officer or from the Fire and Rescue Service prior to any planning application being submitted for approval.

If the Fire and Rescue Service is approached directly and they duly provide advice (as required by them under the *Fire and Rescue Services Act* 2004) the Fire and Rescue Service response will be of an informal nature and not legislatively enforceable unless pertaining to the RR(FS)O once the building is occupied. The advice tendered will normally be in writing and will clearly define which contents are informal in nature, and which are given to meet the requirements of the Order.

Part 2, section 6 of the *Fire and Rescue Services Act* 2004 states:

6. Fire safety

(1) A fire and rescue authority must make provision for the purpose of promoting fire safety in its area.

(2) In making provision under subsection (1) a fire and rescue authority must in particular, to the extent that it considers it reasonable to do so, make arrangements for—

(a) the provision of information, publicity and encouragement in respect of the steps to be taken to prevent fires and death or injury by fire;

(b) the giving of advice, on request, about—

(i) how to prevent fires and restrict their spread in buildings and other property;

(ii) the means of escape from buildings and other property in case of fire.

It is recommended that the designer's first approach be made to Building Control, who will be able to advise which other interested parties should be

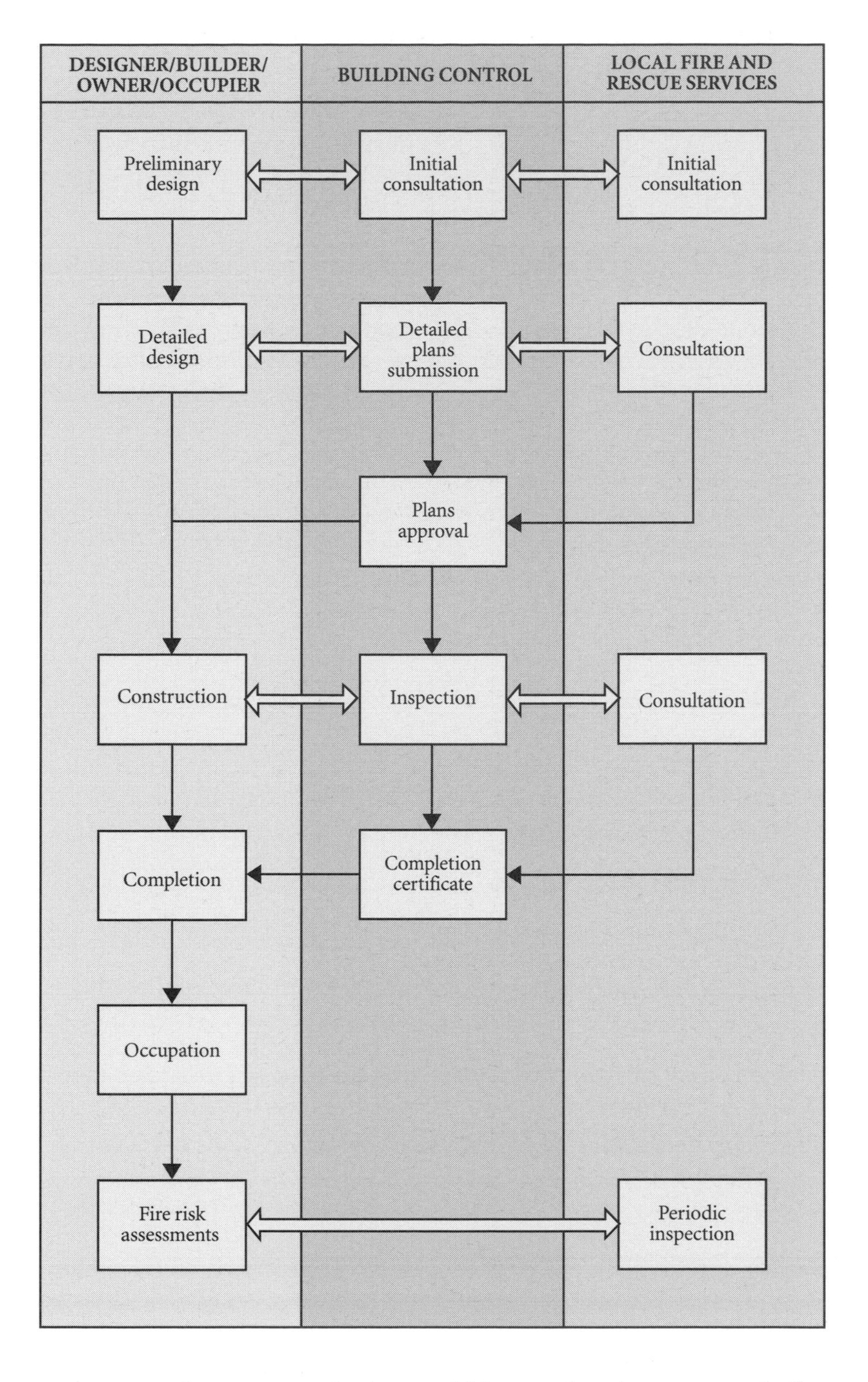

involved in any current or future consultations relating to the project.

It is at this initial stage in the design of a building/development that fire engineered solutions should be discussed in principle where they are proposed as alternatives to the ADB approach to Building Regulations compliance.

There is an increasing trend for architects and designers to move away from the recommendations contained within ADB in order to maximise the potential of their building or development. The possible benefits to be gained from adopting an alternative approach to Building Regulations compliance range from the purely aesthetic (for example, retaining a maximum open aspect to an expansive building) to substantial construction cost savings (this can be demonstrated, for example, by a reduction of required structural steel fire protection over the prescriptive guidance in ADB).

With an increasing use of Fire Engineering within the building and construction industries, the role of Building Control Officers has become more complex than ever; simple knowledge and application of the recommendations contained within ADB (produced in support of the Building Regulations) is no longer considered sufficient.

ADB provides a level of guidance which when followed to the letter should guarantee Building Regulations compliance in so far as fire safety is concerned. It is not, however, the panacea for fire and life safety. At the very beginning of the document the following text appears:

> *The Approved Documents are intended to provide guidance for some of the more common building situations. However, there may well be alternative ways of achieving compliance with the requirements.*
>
> ***Thus there is no obligation to adopt any particular solution contained in an Approved Document if you prefer to meet the relevant requirement in some other way.***

Note: 'the requirements' in this instance refers to the *Building Regulations* 2000.

This flexible approach is further supported by the contents of numerous Home Office circulars which since 2000 have advised Building Control Departments and Fire and Rescue Services alike that the application of the guidance in ADB is not the only method of achieving compliance with the Building Regulations.

There are a number of publications available which provide the designer with alternative ways to meet the requirements of the Building Regulations (BS5588 series and BS7974 series to quote but two examples). A designer is not restricted to employing methods described in any of the guidance documentation, but it is often much easier to justify an engineered solution if he or she does so.

Example

1,000,000 sq ft development in London

Fire engineering provided the following solutions:

- removed two protected means of escape stairs by evacuation assessment;
- removed pressurisation to ten staircases;
- reduced fire rating between buildings from 120 to 90 minutes;
- negated the requirement for fire rated glazing in the atria;
- bespoke fire detection and alarm system;
- highly phased evacuation system to minimise disruption to operations;
- omitted the inclusion of hose reels throughout the building.

And so the problem arises – if the plans submitted for approval do not follow the guidance contained in ADB, who determines whether the design is 'fit for purpose' and complies with the requirements of the Building Regulations?

Whilst fire engineered solutions permit freedom for design they also present one serious problem; relatively few people have sufficient understanding of fire science and engineering to readily appreciate the methodology applied in an engineered solution.

The Fire and Rescue Services, together with Building Control have to understand the theory behind any fire engineered solution, its impact on the future occupants of the building and any burden it may place on the management of the building.

In reaching a decision, either or both authorities may seek advice from other parties such as the Local Fire Engineering Group of the Fire and Rescue Services or independent fire safety consultants.

In order to present a case for a fire engineered solution, the designer must produce a fire strategy for the building which describes in detail how the building design (including any Fire Engineering) achieves compliance with the Building Regulations. This documentation will form part of the planning approval and should be handed over to the eventual responsible person for the premises to be used in their fire risk assessment process. Without this explanatory documentation it is highly likely that the Building Control Officer and the Fire and Rescue Service will simply reject the submitted plans as non-compliant.

As part of the Fire and Rescue Service restructuring, a number of Fire Engineering Groups have been set up around the country within the service, with the primary objective of providing fire engineering advice; in the event that a planning application contains proposals which do not follow the guidance of ADB (as the previous examples), the Building Control Officer and/or Fire and Rescue Service will seek detailed advice from their local Fire Engineering Group.

During the detailed design stage of a project there is a statutory requirement for consultation with the enforcing authority if the eventual occupation or use of the building falls within the remit of the RR(FS)O. The Building Control Officer will be responsible for assessing the design in accordance with the requirements of the Building Regulations, whereas the Fire and Rescue Service assume the role of ensuring compliance with the order, including looking forward to when the building is put to its intended use.

The final round of consultations involving the Building Control Officer and Fire and Rescue Service occurs prior to the issuing of a completion notice for the premises. It is quite possible that the premises may become partly or wholly occupied before the issue of a completion notice; this being the intended case the Building Control Officer must be informed of the date and extent of the intended occupation – giving (currently) at least five days' notice of such an event.

Once the premises are occupied, either in whole or in part, the Fire and Rescue Service become directly involved in the fire safety of the relevant persons and will commence their own assessment (if not already done) to determine the potential risk of occupation and the requirement for (and frequency of if so required) RR(FS)O inspections.

THE FIRE SERVICE AND OCCUPIED PREMISES

The Fire and Rescue Service will usually carry out the general role of enforcing authority in so far as fire safety and compliance with the RR(FS)O is concerned. (But also refer to the chart at the beginning of this chapter for some exceptions).

The Fire Service has adopted an 'Enforcement Management Model' (EMM) produced by the HSE for performing fire safety audits on premises. This model is intended to provide a consistent base from which to make assessments and provide an auditable trail should the result of an assessment be challenged.
The EMM considers and provides base guidance for numerous aspects of fire safety including:

- relative risk based on premises type;
- risk factor determined by premises size;
- responsible person factors;
- strategic factors.

In addition the EMM uses a number of decision tree flow charts to ensure consistency of approach to all audits.

The premises can then, after audit, be classified into one of a number of compliance levels, which will determine the expected enforcement action, if any, to be taken and may range from 'educate and inform' at one end of the scale, to 'prohibition' at the other.

It must be noted that irrespective of the overall compliance level of the building, prosecution and/or prohibition could still be considered depending upon the circumstances of each individual case.

Inspections will be made of premises deemed by the enforcing authority as posing a credible risk to its occupants and/or the immediate community; to this end, a system of prioritisation will be employed, this is discussed further in the next section of this chapter.

Enforcement activities will be carried under the auspices of the Integrated Risk Management Planning (IRMP), part of the modernisation process of the Fire and Rescue Services.

If, as a result of an inspection of premises, the enforcing authority has any concerns whatsoever with the fire safety provisions provided there is recourse to the serving of notices on the responsible person/s.

The serving of a notice on a responsible person in respect to premises does not necessarily entail the immediate cessation of operations or occupation; the notices and their effects are discussed in a later section of this chapter.

FIRE SERVICE PRIORITISATION OF PREMISES

As part of the recent Fire Service shake-up, the concept of IRMP was introduced. This enables Fire Services to generate a 'risk ranking' table for the different types of premises within their area.

The IRMP National Document provides Fire Services with a framework on which to build the 'business' of fire fighting and take it into the future. Risk management involves the assessment of any perceived risk and the provision of an accountable response to the risk; this approach has been adopted by Fire Services in the demographic categorisation of areas based on perceived risk levels and recorded historical data. The categorisation provides a bench-mark for improvement in those areas deemed to have a higher than average risk from fire for the area served, and also provides an indication of the level of turn-out required for attendance at a fire incident.

Fire Precautions Act 1971: Circular 29 provides guidance from which a Fire Service can develop a detailed premises risk table. The Circular, by necessity, groups premises into a number of applicable groups (17 in total) for the purposes of apportioning a risk factor. In reality, there are many more than 17 premises types and to this end Appendix B to the Circular provides a comparison of those referred to in the Annual Fire Safety Returns and the 17 in the Circular.

Although providing a starting point for premises classification, the 'pigeon holing' of premises into a small number of types is largely open to the judgment of the individual compiling the classification.

The Chief Fire Officers Association (CFOA) has issued further detailed tables comparing different types of occupancy, their grouping according to *Fire Precautions Act* 1971: Circular 29, and an initial median risk ranking for each.

The following table is a conglomeration of information contained in the CFOA tables and that of Circular 29, with a small number of additions and amendments.

The risk factors listed in the table are median values from a range of possible values for each premises type and are expected to be the basis upon which the Fire Service begin their assessment of occupancies in their area. The values given are for the purposes of comparison between the premises types and relate to the 'overall potential for loss of life, or serious injury'. The information supplied below was correct at the time of collection but may be subject to change in the future.

It is likely therefore that in prioritising their inspection plan, the Fire Service will look at the overall risk presented by the premises (and to whether the premises would have previously held a fire certificate) in deciding which should be 'top of the list' for inspection.

Description	FSA Circular 29 group	Risk factor
HMO	HMO	6
Hospital (NHS)	Hospital	6
Hospital (Private)	Hospital	6
Hostel	HMO	6
Accommodation for students under 18 by further education colleges	Care home	5
Adult placement schemes	Care home	5
Ambulance station (with sleeping accommodation)	Other sleeping accommodation	5
Betting shop	Shop	5
Bingo hall/licensed	Licensed premises	5
Bingo hall/non licensed	Public building	5
Boarding schools	Care home	5
Camping site	Other sleeping accommodation	5
Car showroom	Shop	5
Caravan park	Other sleeping accommodation	5
Care home for adult placements	Care home	5
Care home for older people	Care home	5
Casino	Licensed premises	5
Chalet park	Other sleeping accommodation	5
Childminders	Care home	5
Children's homes	Care home	5
Cinema	Licensed premises	5
Club (social)	Licensed premises	5
Club house (licensed) (members)	Licensed premises	5
Day care providers	Care home	5
Domiciliary care	Care home	5
Fire station (with sleeping accommodation)	Other sleeping accommodation	5
Food court	Shop	5
Foster homes	Care home	5
Guesthouse	Hotel	5
Hairdressing salon	Shop	5
Holiday centre/hotel	Hotel	5
Holiday centre/licensed	Hotel	5
Holiday centre/other sleeping accommodation	Hotel	5
Hotel	Hotel	5
Houses converted to flats three floors and over	Houses converted to flats	5
Houses converted to flats up to two floors	Houses converted to flats	5
Hypermarket	Shop	5
Kiosk/office	Shop	5
Kiosk/shop/factory	Shop	5
Laundrette	Shop	5
Licensed – cafe	Licensed premises	5
Licensed – restaurant	Licensed premises	5
Market (indoor)	Shop	5
Military barracks	Other sleeping accommodation	5
Motel	Hotel	5
Nightclub	Licensed premises	5
Petrol filling station	Shop	5
Police station (with cells)	Other sleeping accommodation	5
Post office	Shop	5
Prison (with cells)	Other sleeping accommodation	5
Public house	Licensed premises	5
Residential family centres	Care home	5
Residential school, college or university	Other sleeping accommodation	5
Residential special schools	Care home	5
Royal palaces, other Crown with sleeping accommodation	Other sleeping accommodation	5
Shop not listed below	Shop	5
Shopping centre	Shop	5
Showroom	Shop	5
Superstore	Shop	5
Theatre	Licensed premises	5
Unlicensed – cafe	Shop	5
Unlicensed – restaurant	Shop	5
Wine bar	Licensed premises	5
Airport or ferry terminal	Other premises open to the public	4

Description	FSA Circular 29 group	Risk factor
Ambulance station (no sleeping accommodation)	Other workplaces	4
Amusement arcade	Other premises open to the public	4
Amusement park	Other premises open to the public	4
Animal boarding or breeding establishment	Other workplaces	4
Bank	Office	4
Brickworks	Factories and warehouses	4
Bus station	Other premises open to the public	4
Business unit	Other workplaces	4
Club house (not licensed)	Public building	4
College	School	4
College	Further education	4
Community centre	Public building	4
Computer centre	Office	4
Crematorium	Other premises open to the public	4
Day nursery	School	4
Explosive or highly-flammables store	Factories and warehouses	4
Factory	Factories and warehouses	4
Fire station (no sleeping accommodation)	Other workplaces	4
Football ground	Other premises open to the public	4
Forces careers office	Office	4
Garage	Other workplaces	4
Golf course	Other premises open to the public	4
Hall	Public building	4
Health centre	Other premises open to the public	4
Industrial miscellaneous	Other workplaces	4
Laboratories/research establishment	Other workplaces	4
Law court	Other premises open to the public	4
Leisure centre	Public building	4
Library	Public building	4
Mill	Factories and warehouses	4
Museum	Public building	4
Office	Office	4
Offices (local government)	Office	4
Offices (local government)/open to the public	Other premises open to the public	4
Other education, training and culture	Further education	4
Other industrial mineral (structure)	Other workplaces	4
Outdoor public event	Other premises open to the public	4
Place of worship	Other premises open to the public	4
Police station (no cells)	Office	4
Private car park (covered or multi-storey)	Factories and warehouses	4
Public car park (covered or multi-storey)	Other premises open to the public	4
Quarry	Other workplaces	4
Railway premises (not stations)	Other workplaces	4
Railway station	Other premises open to the public	4
Retail warehouse	Factories and warehouses	4
School	School	4
Sewage treatment works (structure)	Other workplaces	4
Sports centre	Public building	4
Sports ground	Other premises open to the public	4
Stadium	Other premises open to the public	4
Storage depot/factory or warehouse	Factories and warehouses	4
Storage depot/other workplace	Factories and warehouses	4
Store	Factories and warehouses	4
Sub-surface railway station	Other premises open to the public	4
Surgery	Other premises open to the public	4
Swimming pool	Public building	4
Tennis centre	Other premises open to the public	4
University	Further education	4
Vacant or unoccupied premises	Other workplaces	4
Vehicle repair	Other workplaces	4
Warehouse	Factories and warehouses	4
Works	Factories and warehouses	4
Workshop	Other workplaces	4

This table is presented purely as a guide to the likely priority certain premises may be given by the Fire Service when deciding which to inspect first; the data should not be used in the production or carrying out of a fire risk assessment as required by the RR(FS)O.

The application of risk identification and management to premises is the key to RR(FS)O compliance. As the responsible person, and let's assume you are if you have read this far, you are obliged to carry out (or have carried out on your behalf) a fire risk assessment for your premises. The Fire Service will inspect your premises and provide fire safety advice, but they will not carry out your fire risk assessment.

A fire service premises inspection need not be (and indeed is not intended to be) a traumatic experience, if you have followed the guidance in this book and the DCLG guides for premises any visit from the fire service should be a straightforward procedure. The fire service is not a draconian institute, you will not be subjected to a 'Spanish inquisition' style grilling, nor will you be expected to know everything there is to know about fire safety, fire science, fire engineering, etc. What you must know, and be able to demonstrate, is that you are aware of the fire safety measures in place on your premises and their suitability and applicability to the occupants and visitors thereto.

If you follow the guidance in the DCLG guides for your premises you will (assuming reasonable records are maintained) be in a position to provide the fire service inspector with everything he needs for his assessment visit. You may be requested to provide copies of any fire safety related documentation kept for the fire service inspector to take away – it is your duty to provide this information if so requested (yet another good reason for a documented risk assessment, irrespective of premises or organisational size).

Fire service officers have a duty to ensure that you are doing everything *practicable* to maximise the safety of people on or around your premises (and that includes you!) in the event of a fire.

The philosophy of, and a suggested methodology for, fire risk assessment are included in Chapter 4 of this book.

In a nutshell:

- Building control duty to consult with LFRS.
- LFRS prioritisation of premises for inspection.
- LFRS likely to be the enforcing authority for occupied premises.

NOTICES

In the event that the enforcing authority deems that the premises present a risk to occupants, or could present a risk to relevant persons, it has the power to issue a notice on the responsible person for the premises depending upon how severe the risk is perceived to be.

The RR(FS)O makes provision for three notice levels:

- Alterations Notice;
- Enforcement Notice;
- Prohibition Notice.

Alterations notice

An alterations notice may be served on a responsible person if the enforcing authority is of the opinion that the premises, their use or contents, pose a risk to relevant persons; or, where a change of use, occupation or process would pose a risk to relevant persons. This notice effectively replaces the old fire certificate scheme for certain premises where the nature of the premises, occupants or the use to which the premises are put, pose a real risk to relevant persons.

Example

Petrochemical plant: an alterations notice may be served which specifies the processes, materials or activities which pose the risk to persons, or those which may constitute a risk to persons in the event of any operational changes being made. Any specific measures put in place to maximise the safety of relevant persons are also recorded on the notice.

An alterations notice does not prevent the occupation or use of the premises but does require that any subsequent changes which depart from or affect the contents of the notice are, prior to implementation, the subject of a documented risk assessment and notification of the change must be made to the enforcing authority.

Example

A pharmaceuticals plant adds a secondary distillation process to the existing plant. This addition increases the number of potential points of release of a flammable material. The change to the process plant materially changes the existing risk to relevant persons and the plant fire risk assessment must be revisited and the enforcing authority notified.

Example of an Alterations Notice

Dear Sirs

THE REGULATORY REFORM (FIRE SAFETY) ORDER 2005

{Address}

The Fire and Rescue Authority are the enforcing authority, under Article 25 of the above legislation.

The attached Alterations Notice is a legal requirement to which you have a right of appeal to a Magistrates' Court. Your attention is drawn to the notes that accompany this Notice.

If you are in any doubt as to the obligations placed upon you, or there is any relevant matter upon which you require clarification, you may write to me direct, or alternatively contact the inspector named above.

Yours faithfully

for Chief Officer

Enc:
Alterations Notice
Notes and Standard Terms and Definitions

{ xxxxxx FIRE AND RESCUE SERVICE }
ALTERATIONS NOTICE

NOTICE REQUIRING STEPS TO BE TAKEN UNDER
ARTICLE 29 OF THE REGULATORY REFORM (FIRE SAFETY) ORDER 2005

Name: { }

Premises: { }

Address: { }

I {xxxxxx} on behalf of {xxxxxx Fire and Rescue Authority), hereby give you notice that the Fire and Rescue Authority are of the opinion in respect to the above named premises, that the premises constitute a serious risk to relevant persons (see notes) due to {xxxxxx}.

I {xxxxxx} on behalf of {xxxxxx), hereby give you notice that the Fire and Rescue Authority are of the opinion that any change made to the premises, or the use to which they are put, may constitute a serious risk to relevant persons (see notes) due to {xxxxxx}.

The Fire and Rescue Authority hereby direct that if you, as a responsible person (see notes), intend making any of the following –

a a change to the premises;
b a change to the services, fittings or equipment in or on the premises;
c an increase in the quantities of dangerous substances which are in or on the premises, or
d a change to the use of the premises;

which may result in a significant increase in risk, you as the responsible person (see notes) must notify the Fire and Rescue Authority of the proposed changes.

The Fire and Rescue Authority also direct that in addition to the notification referred to above, that as a responsible person you must –

a take all reasonable steps to notify the terms of this notice to any other person, or persons, who have to any extent control of the premises, insofar as the requirements in Articles 8 to 22 of the Regulatory Reform (Fire Safety) Order 2005, or in regulations made under Article 24, relates to matters under his, or their, control;
b carry out or review the risk assessment and record the significant findings, including the measures which have been taken or will be taken and identify any group of persons identified by the risk assessment as being especially at risk;
c record the arrangements as are appropriate, having regard to the size of his undertaking and the nature of its activities, for the effective planning, organisation, control, monitoring and review of the preventative and protective measures, and
d before making any changes referred to in the above paragraph, send to the Fire and Rescue Authority a copy of the risk assessment and summary of the changes proposed to be made to the existing general fire precautions.

This Notice shall be deemed to be in force until such time as it is withdrawn by the Fire Authority or cancelled by the court. You have a right to appeal against this Notice (see notes), by way of complaint for an order, to the Clerk to the Court of the Magistrates' Court acting for the petty sessions area in which your premises is located. If you wish to appeal, you must do so within 21 days of the date of this notice.

Date: ....   Signed: ....
(On behalf of and duly authorised by the Fire Authority)

NOTES TO ACCOMPANY ALTERATIONS NOTICE SERVED UNDER ARTICLE 29 OF THE REGULATORY REFORM (FIRE SAFETY) ORDER 2005

1 Contravention of any requirement imposed by an alterations notice is an offence under Article 32 of The Regulatory Reform (Fire Safety) Order 2005 and renders the offender liable, on summary conviction, to a fine not exceeding the statutory maximum or, on conviction on indictment, to an unlimited fine, or imprisonment for a term not exceeding two years, or both.

2 In any proceedings for an offence referred to in Note 1. Where the commission by any person of an offence under the Order, is due to the act or default of some other person, that person is guilty of the offence, and a person may be charged with and convicted of the offence whether or not proceedings are taken against the first mentioned person.

3 Nothing in the Order operates so as to afford an employer a defence in any criminal proceedings for a contravention of those provisions by reason of any act or default of an employee or person nominated to implement measures for fire-fighting and procedures for serious and imminent danger and for danger areas, and appointed to assist him/her in undertaking such preventive and protective measures as necessary.

4 Subject to Note 3, in any proceedings for an offence under the Order, except for a failure to comply with Articles 8(a) (Duty to take general fire precautions) or 12 (Elimination or reduction of risks from dangerous substances), it is a defence for the person charged to prove that he/she took all reasonable precautions and exercised due diligence to avoid the commission of such an offence.

5 In any proceedings for an offence under the Order consisting of a failure to comply with a duty or requirement so far as is reasonably practicable, it is for the accused to prove that it was not reasonably practicable to do more than was in fact done to satisfy the duty or requirement.

6 A person on whom an alteration notice is served may under Article 35 of the Order appeal to the Magistrates Court within 21 days from the date on which the alterations notice was served. The bringing of an appeal has the effect of suspending the operation of the notice until the appeal is finally disposed of or, if the appeal is withdrawn, until the withdrawal of the appeal.

7 To satisfy the 'Environment and Safety Information Act 1988' the Fire and Rescue Authority is obliged to enter details of any alterations notice into a register to which the public have access. If you feel that any such entry would disclose secret or confidential trade or manufacturing information then you should appeal in writing to the Fire and Rescue Authority within a period of 14 days following the service of the Notice.

8 The requirements in the Notice are only intended to maintain the current level of fire precautions whilst proposed changes and possible consequences are considered by the Fire and Rescue Authority.

9 The Notice is issued without prejudice to any other enforcement action that may be taken by this or any other enforcement authority.

Enforcement notice

An enforcement notice may be served on any premises (or processes therein) that the enforcing authority deems not to comply with the requirements of the RR(FS)O.

> **Example**
>
> Upon inspection by the enforcing authority, a newly refurbished office is discovered to have defective emergency lighting and insufficient portable fire extinguishers for the size and type of establishment. The enforcing authority serves upon the responsible person an enforcement notice listing the areas of non-compliance with the RR(FS)O and stating that corrective action must be carried out to the satisfaction of the enforcing authority within 60 days.

An enforcement notice does not prevent the occupation or use of the premises but documents any observations made by the enforcing authority in relation to the premises and non-compliance with the RR(FS)O, and requires that the responsible person take action to remedy the non-compliances within a stated period of time (in any case not less than 28 days). The enforcement notice *may* include remedial actions to be taken to remedy any areas of non-compliance but must only offer these actions as guidance and not as the sole means by which compliance can be achieved.

An enforcement notice may be withdrawn by the enforcing authority at any time, usually as the result of a further inspection of the premises to which the issued notice applies.

> **Example**
>
> The office mentioned in the previous example remains occupied but the responsible person takes the necessary actions to correct the non-compliant issues. The enforcing authority re-inspects the premises and deems the actions taken to be sufficient to comply with the requirements of the RR(FS)O. The enforcement notice is withdrawn.

Example of an Enforcement Notice

Recorded Delivery

Dear Sirs

THE REGULATORY REFORM (FIRE SAFETY) ORDER 2005

{xxxxxx}

The (Fire and Rescue Authority – insert details) are the enforcing authority, under Article 25 of the above legislation.

Following a fire safety audit of the above premises by one of my inspectors on {xxxxxx} I confirm that the matters and steps specified on the attached Schedule need to be carried out to remedy your failure to comply with the above legislation.

The attached Enforcement Notice is a legal requirement to which you have a right of appeal to a Magistrates' Court. Your attention is drawn to the notes that accompany this Notice.

If you are in any doubt as to the obligations placed upon you, or there is any relevant matter upon which you require clarification, you may write to me direct, or alternatively telephone your enquiry to the inspector named above.

Yours faithfully

for Chief Officer

Enc:
Enforcement Notice
Schedule to Enforcement Notice
Notes and Standard Terms and Definitions

{ xxxxxx FIRE AND RESCUE SERVICE }
ENFORCEMENT NOTICE

NOTICE REQUIRING STEPS TO BE TAKEN UNDER ARTICLE 30 OF THE REGULATORY REFORM (FIRE SAFETY) ORDER 2005

Name: { }

Premises: { }

Address: { }

I {xxxxxx}on behalf of the (xxxxxx Fire and Rescue Authority), hereby give you notice that the Fire and Rescue Authority are of the opinion that, as a person being under an obligation to do so, you have failed to comply with the requirements placed upon you by The Regulatory Reform (Fire Safety) Order 2005 in respect of the above named premises and the persons who may be on the premises or who may be affected by a fire on the premises.

The provisions of the regulatory Reform (Fire Safety) Order 2005 which have not been complied with are:

{xxxxxx}

The matters which, in the opinion of the Fire and Rescue Authority, result in the failure to comply with the aforementioned provisions of the Regulatory Reform (Fire Safety) Order 2005 are specified in the Schedule to this Notice.

The Fire and Rescue Authority are further of the opinion that the steps identified in the Schedule to this Notice must be taken to remedy the specified failure(s) to comply with The Regulatory Reform (Fire Safety) Order 2005.

Unless the steps identified in the Schedule to this Notice have been complied with by {specify date} you will be regarded as not being in compliance with this Notice and the Fire and Rescue Authority may consider a prosecution against you. You may, however, apply for an extension to this time limit (see notes).

You have the right to appeal against this notice (see notes), by way of complaint for an order, to the Clerk to the Court of the Magistrates' Court acting for the petty sessions area in which your premises is located. If you wish to bring an appeal, you must do so within 21 days of the date this notice is served on you. The Magistrates' Courts Act 1980 will apply to the proceedings. The bringing of an appeal shall have the effect of suspending the operation of this Enforcement Notice until the appeal is finally disposed of or, if the appeal is withdrawn, until the withdrawal of the appeal.

Date: .... Signed: ....
(On behalf of and duly authorised by the Fire and Rescue Authority)

SCHEDULE REFERRED TO IN ENFORCEMENT NOTICE NO {xxxxxx} REQUIRING STEPS TO BE TAKEN UNDER ARTICLE 30 OF THE REGULATORY REFORM (FIRE SAFETY) ORDER 2005 ISSUED BY THE {xxxxxx} FIRE AND RESCUE AUTHORITY ON {xxxxxx}

Name and Address of Premises:

Where appropriate, a plan may form part of this Schedule to illustrate the steps which, in the opinion of the Fire and Rescue Authority, need to be taken in order to comply with The Regulatory Reform (Fire Safety) Order 2005. Note: Notwithstanding any consultation undertaken by the Fire and Rescue Authority, **before** you make any alterations to the premises, **you** may need to apply for approval from either the Local Authority Building Control or an Approved Inspector and/or the approval of any other bodies having a statutory interest in the workplace.

SCHEDULE

The location and details of matters which are considered to be failures to comply with The Regulatory Reform (Fire Safety) Order 2005 are detailed in the column below.	**The steps considered necessary to remedy the failures are detailed in the column below.**

NOTES TO ACCOMPANY ENFORCEMENT NOTICE SERVED UNDER ARTICLE 30 OF THE REGULATORY REFORM (FIRE SAFETY) ORDER 2005

1 Application to premises. The Regulatory Reform (Fire Safety) Order 2005, subject to paragraphs 6(1)(a)–(g) of the RR(FS)O listed below, applies to any premises.

2 The Order does not apply in relation to—

(a) domestic premises;
Note: Where the premises are, or consist of, a house in multiple occupation this Order applies in relation to those parts of the premises which are not domestic premises.
(b) an offshore installation within the meaning of Regulation 3 of the Offshore Installation and Pipeline Works (Management and Administration) Regulations 1995;
(c) a ship, in respect of the normal ship-board activities of a ship's crew which are carried out solely by the crew under the direction of the master;
(d) fields, woods or other land forming part of an agricultural or forestry undertaking but which is not inside a building and is situated away from the undertaking's main buildings;
(e) an aircraft, locomotive or rolling stock, trailer or semi-trailer used as a means of transport or a vehicle for which a licence is in force under the Vehicle Excise and Registration Act 1994 or a vehicle exempted from duty under that Act;
(f) a mine within the meaning of Section 180 of the Mines and Quarries Act 1954, other than any building on the surface at a mine, and
(g) a borehole site to which the Borehole Sites and Regulations 1995 apply.

3 You may appeal (under Article 35 of the Order) against an enforcement notice served (under Article 30 of the Order). The appeal is made, within 21 days from the day on which the notice is served, to the Magistrates' Court for the area in which your premises is situated and may be brought on the grounds that you think that:

(a) the service of an enforcement notice was based on an error of fact;
(b) the service of the enforcement notice was wrong in law, and
(c) the Fire and Rescue Authority erred in the exercise of their discretion in serving the enforcement notice.

Without prejudice to the breadth of the grounds of appeal set out in paragraphs (a) to (c) above, examples of situations in which an appeal may lie are where:

(a) you dispute any of the facts in the Notice which detail the steps which have to be taken in order to comply with any provision of the Order,
(b) you think that an unreasonable time period has been set for the taking of the steps set out in the Notice.

4 The Fire and Rescue Authority may grant, at their discretion, an extension (or further extension) of time specified for the steps to be taken if an appeal against the Notice is not pending. Application for an extension of time should be addressed to: The Chief Officer {name and address of Fire and Rescue Authority}.

5 Failure to comply with any requirement imposed by an enforcement notice served under Article 30 of the Order within the time specified in the Notice (or such further time as the Fire and Rescue Authority may, at their discretion, grant) is a criminal offence under Article 32(d) of the Order. A person guilty of such an offence shall be liable:

(a) on summary conviction to a fine not exceeding the statutory maximum, or
(b) on conviction on indictment, to a fine or to imprisonment for a term not exceeding two years, or both.

6 In any proceedings for an offence referred to above, where the commission by any person of an offence under the Order, is due to the act or default of some other person, that person is guilty of the offence, and a person may be charged with and convicted of the offence whether or not proceedings are taken against the first mentioned person.

7 Nothing in the Order operates so as to afford an employer a defence in any criminal proceedings for a contravention of those provisions by reason of any act or default of an employee or person nominated to implement measures for fire-fighting and procedures for serious and imminent danger and for danger areas, and appointed to assist him/her in undertaking such preventive and protective measures as necessary.

8 Subject to Note 9, in any proceedings for an offence under the Order, except for a failure to comply with Articles 8(1) (Duty to take general fire precautions) or 12 (Elimination or reduction of risks from dangerous substances), it is a defence for the person charged to prove that he/she took all reasonable precautions and exercised due diligence to avoid the commission of such an offence.

9 If you are the responsible person you are under an obligation to comply with the provisions of the Order or of any regulations made under it. If you have failed to comply and you and the Fire and Rescue Authority cannot agree on the measures which are necessary to remedy the failures(s). Under Article 36 of the Order you and the Fire and Rescue Authority may agree to refer the question, as to what measures are necessary to remedy the failure(s), to the Secretary of State for a determination.

10 It should be noted that in order to satisfy the 'Environment and Safety Information Act 1988' the Fire and Rescue Authority is obliged to enter details of any enforcement notice into a register to which the public have access. If you feel that any such entry would disclose secret or confidential trade or manufacturing information then you should appeal in writing to the Authority within a period of 14 days following the service of the Notice.

11 To assist with administration procedures, it would be helpful if you could quote the reference number (at the top of this notice) when dealing with the Fire and Rescue Authority.

STANDARD TERMS AND DEFINITIONS

FIRE RESISTING (FIRE RESISTANCE): The ability of a component or construction of a building to satisfy, for a stated period of time, some or all of the appropriate criteria specified in the relevant British Standard.

INTUMESCENT STRIPS: A strip of material placed along the door edges (excluding the bottom edge), or frame, that will react to heat by expanding to form a seal to the passage of hot gases and flame.

SMOKE SEAL: A flexible strip of material (often used in conjunction with an intumescent strip) placed along the door edges, or frame, to limit the spread of cold smoke during the early stages of a fire.

SELF-CLOSING DEVICE: A device which is capable of closing the door from any angle and against any latch fitted to the door. Rising butt hinges are not acceptable.

AUTOMATIC DOOR RELEASE: A device, linked to (or operated by the sound of) the fire alarm system, that when fitted to a fire resisting self-closing door, enables it to be held open during normal working conditions.

EMERGENCY ESCAPE LIGHTING: That part of the emergency lighting system provided for use when the electricity supply to the normal lighting fails so as to ensure that the means of escape can be safely and effectively used at all times.

RISK ASSESSMENT: An organised appraisal of your activities and premises to enable you to identify potential fire hazards, and to decide who (including employees and visitors) might be in danger in the event of fire, and their location. You will then evaluate the risks arising from the hazards and decide whether the existing fire precautions are adequate, or whether more needs to be done. It will be necessary for you to record your findings (if you have five or more employees), and to review and revise when necessary.

Prohibition notice

A prohibition notice may be served by the enforcing authority where it believes that the premises, or their use, involve a clear and present risk to relevant persons to an extent that such use should be restricted or prohibited.
A prohibition notice is the most onerous notice which can be served by the enforcing authority and may effectively require the immediate cessation of any activities or processes listed in the notice as presenting an imminent and serious threat to the safety of relevant persons.

The prohibition notice must state:

- the matters which may give rise to the unacceptable risk;
- that the use to which the premises are put, as described above is prohibited or restricted until remedied.

A prohibition notice *may* contain direction for remedial actions to be taken to rectify matters in contravention of the RR(FS)O but must be written in such a way as to allow the responsible person flexibility in adopting alternative solutions.

The notice takes effect immediately if, in the opinion of the enforcing authority, the risk of serious personal injury is clear and already present, or would become imminent should the premises be put to their intended use without further addressing the issue of fire safety.

A prohibition notice may be withdrawn by the enforcing authority at any time.

Example of a Prohibition Notice

Dear Sirs

REGULATORY REFORM (FIRE SAFETY) ORDER 2005, ARTICLE 31

{xxxxxx}

Please find herewith a Prohibition Notice in respect of the above premises.

Your attention is drawn to the notes which accompany this Notice.

Any queries concerning these matters may be directed to the inspector named above but all correspondence should be addressed to me.

Yours faithfully

for Chief Officer

Enc:
Prohibition Notice
Notes to Accompany Prohibition Notice

{ xxxxxx FIRE AND RESCUE SERVICE }
REGULATORY REFORM (FIRE SAFETY) ORDER 2005 ARTICLE 31
PROHIBITION NOTICE

Name: {xxxxxx}

Address: {xxxxxx}

Occupier of: {xxxxxx}

I {xxxxxx} on behalf of the (xxxxxx Fire and Rescue Authority) hereby give you notice that the Fire and Rescue Authority are of the opinion that the following use of the said premises as occupied by you, namely {xxxxxx} involves, or will involve, a risk to persons on the premises in the event of fire so serious that use of the premises ought to be prohibited or restricted.

The Fire and Rescue Authority are further of the opinion that the matter(s) which give rise/will give rise* to the said risk is/are*:

{ xxxxxx}

and the Fire and Rescue Authority hereby direct that the use of the premises to which this Notice relates is prohibited or restricted as follows:

{xxxxxx}

until the matters specified above have been remedied.

The Fire and Rescue Authority are of the opinion that the risk of injury is/would* be imminent and the prohibition/restriction is to take effect immediately/from*
{xxxxxx}.

This Notice continues in force until the specified matters have been remedied and the Fire and Rescue Authority withdraw it.

*delete as appropriate

Signature: .... Date: ....
(Being a person authorised by {xxxxxx} Fire and Rescue Authority under Section 101 of the Local Government Act 1972 to issue such a notice.)

NOTES TO ACCOMPANY PROHIBITION NOTICE SERVED UNDER ARTICLE 31 OF THE REGULATORY REFORM (FIRE SAFETY) ORDER 2005

1. Under Article 32(2)(h) of this Order it is an offence for any person to fail to comply with any prohibition or restriction imposed by a Prohibition Notice and renders the offender liable, on summary conviction, to a fine not exceeding the statutory maximum or, on conviction on indictment, to a fine, or to imprisonment for a term not exceeding two years, or to both.

2. In any proceedings for an offence referred to in Note 1. Where the commission by any person of an offence under the Order, is due to the act or default of some other person, that person is guilty of the offence, and a person may be charged with and convicted of the offence whether or not proceedings are taken against the first mentioned person.

3. Nothing in the Order operates so as to afford an employer a defence in any criminal proceedings for a contravention of those provisions by reason of any act or default of an employee or person nominated to implement measures for fire-fighting and procedures for serious and imminent danger and for danger areas, and appointed to assist him/her in undertaking such preventive and protective measures as necessary.

4. Subject to Note 3, in any proceedings for an offence under the Order, except for a failure to comply with Articles 8(a) (Duty to take general fire precautions) or 12 (Elimination or reduction of risks from dangerous substances), it is a defence for the person charged to prove that he/she took all reasonable precautions and exercised due diligence to avoid the commission of such an offence.

5. In any proceedings for an offence under the Order consisting of a failure to comply with a duty or requirement so far as is reasonably practicable, it is for the accused to prove that it was not reasonably practicable to do more than was in fact done to satisfy the duty or requirement.

6. A person on whom a Prohibition Notice is served may appeal under Article 35 of the said Order to the Magistrates' Court, for the area in which the premises is situated, within 21 days from the date on which the prohibition notice is served. The bringing of an appeal does not have the effect of suspending this Notice, unless on the application of the appellant, the Court so directs (and then only from the giving of the direction).

7. It should be noted that in order to satisfy the 'Environment and Safety Information Act 1988' the Fire and Rescue Authority is obliged to enter details of any prohibition notice into a register to which the public have access. If you feel that any such entry would disclose secret or confidential trade or manufacturing information then you should appeal in writing to the Authority within a period of 14 days following the service of the Notice.

8. The works or actions specified in the Notice are only intended to reduce the excessive risk to a more acceptable level. The Notice is issued without prejudice to any other enforcement action that may be taken by this or any other enforcement authority.

9. The Fire and Rescue Authority would be willing to consider and reasonably assist with any proposals you may have to remedy the matters specified in the Notice.

7 Offences and appeals

OFFENCES

In general non-compliance with any part of the Fire Safety Order constitutes an offence in law.

Key offences may be described as follows.

- Failure to provide adequate fire safety measures as would be expected for the type of premises and use to which they are put. This also encompasses the lack of maintenance of provision of protection for fire-fighting personnel.
- Failure to inform the enforcing authority of a change of use or occupation of premises subject to an alterations notice.
- Failure to comply with any condition or requirement imposed by an enforcement notice.
- Failure to comply with article 23 (General duties of employees at work) where such failure poses a risk of death or serious injury from fire to one or more relevant persons.
- The keeping of a book or record relevant to the RR(FS)O with the knowledge that false or misleading information is contained therein.
- The provision of information relating to the RR(FS)O knowing that information to be false or misleading.
- The wilful obstruction of an inspector in the carrying out of his duties in relation to the RR(FS)O.
- Failure to comply, without reasonable excuse, with any requirement imposed by an inspector in requesting information pertinent to the requirements of the RR(FS)O.
- Impersonation of an inspector.
- The levying of any charge on employees in relation to the imposition of the RR(FS)O.
- Failure to comply with any prohibition order imposed under the RR(FS)O.
- Failure to take such precautions as provisioned by the Secretary of State in relation to the RR(FS)O.
- Failure, without reasonable excuse, to comply with the requirements of article 37 (Luminous tube signs) of the Fire Safety Order.

Any or all of the aforementioned offences may be punishable, upon conviction indictment, by fine and or imprisonment.

DEFENCE

There is very little in the way of defence for not providing adequate fire safety measures

It is, however, a legitimate defence for any person charged under the RR(FS)O that he or she took all reasonable precautions and exercised all due diligence to avoid the commissioning of the associated offence.

The onus, however, is always on the accused to prove that it was not practicable or reasonably practicable to do more than was actually done to satisfy the duty or requirement.

Monetary reasons are rarely considered adequate as defence of an offence in relation to the Order.

APPEALS AND DETERMINATIONS

Within 21 days of any notice being served an appeal against the notice served may be made to a Magistrates' Court. In most cases this will have the effect of suspending the notice until such times as a decision is made by the relevant Court.

In the case of a prohibition notice the lodging of an appeal does not suspend the enforcement of the notice, unless directed by the Court.

If unsatisfied with the outcome of events or the decision of the Magistrates' Court, the person aggrieved has recourse to the Crown Court. Note the aggrieved person can also be the enforcing authority so this is a two-way appeals process.

Determination of disagreements on issues of compliance is made, by agreement of both parties involved, to the Secretary of State. This process already exists within the previous legislative framework and so should require no further explanation here.

The decision of the Secretary of State is final and once determination has been made, the enforcing authority may not take any enforcement action which conflicts with the determination of the Secretary of State, other than if there has been a change to the premises or the use to which they are put, subsequent to the determination.

In a nutshell:

- Onus on the defendant to prove that they took all steps reasonably practicable to ensure the fire safety of the relevant persons of the premises.
- Cost will rarely be accepted as an acceptable defence.
- Appeals can be made via Magistrates' Court, Crown Court and the Secretary of State

8 Miscellaneous provisions

The RR(FS)O makes several provisions under the heading of 'miscellaneous' some of which are briefly outlined below:

Fire-fighters' switches for luminous tube signs, etc.

This re-enforces the requirement for fireman's switches to isolate the low voltage supply to any transformer employed to provide voltages over 'the prescribed voltage' required for luminous signs, etc.

Maintenance of measures provided for protection of fire-fighters

There is now an additional duty on the 'responsible person' to maintain any facilities provided for the fire services, including but not limited to, fire-fighting shafts and lifts, fire service vehicular access to the premises and smoke venting or clearance systems.

Duty not to charge employees for things done or provided

It is not permitted for employers to levy a charge of any kind on their employees in respect of the provision of fire safety measures for the premises, or any other provisions required under the RR(FS)O. This includes any training which may be required in respect of fire safety or fire safety management.

Duty to consult employees

This requirement effectively replaces the existing under the *Fire Precautions (Workplace) Regulations* 1997 with regard to an employer's responsibility to consult with and provide information to their employees with respect to any requirements of the RR(FS)O.

Special provisions in respect of licensed, etc. premises

The relevant licensing authority has a duty to inform the enforcing authority (usually the local Fire Service) of any application for a license, be that entertainment, gambling or other. The licensing authority cannot, however, enforce any terms or conditions to which the RR(FS)O applies – this duty is retained by the enforcing authority.

Suspension of terms and conditions of licenses dealing with same matters as this Order

Any existing terms or conditions imposed by the licensing authority are effectively suspended and have no effect in so far as they relate to the provisions of the RR(FS)O.

Suspension of byelaws dealing with same matters as this Order

Basically, wherever or whenever any byelaw deals with the same matters as the RR(FS)O – i.e. fire safety – they are deemed to have no affect on the premises so long as the premises remain under the umbrella of the RR(FS)O.

Duty to consult enforcing authority before passing plans

When a new development is put before Building Control for planning approval, if the plans are in accordance with the functional requirements of the Building Regulations, the authority has a responsibility to put the said plans and proposals before the enforcing authority prior to passing the plans.

This requirement only applies to plans for premises to which the RR(FS)O applies.

Other consultation by authorities

Should any other authority deem it necessary to take action on a premises which may have an affect under the RR(FS)O they have a duty to consult with the enforcing authority prior to instigating any such action.

Disapplication of the *Health and Safety at Work etc. Act* 1974 in relation to general fire precautions

The *Health and Safety at Work Act* 1974 no longer applies to any premises which are covered by the RR(FS)O, specifically where that Act relates to any matter which could be covered by the RR(FS)O.

Application to the Crown and to the Houses of Parliament

Other than the serving of alteration or enforcement notices, and of course any offences the RR(FS)O will apply to premises owned by or occupied by the Crown; with the exception of the powers of inspectors and the serving of prohibition notices which do not apply when the premises are occupied by the Crown.

Guidance

The Secretary of State (usually via the DCLG or HSE) has a duty to provide guidance to 'responsible persons' relating to their responsibilities in so far as they relate to the RR(FS)O.

Application to visiting forces, etc.

The application of the RR(FS)O to visiting forces applies only in so far as it would be applicable to the Crown as stated above.

Appendix A – UK legislation

The following represents the text of the main body of the RR(FS)O, the five schedules attached to the Order have not been reproduced herein; the Order in its entirety may be obtained free of charge from the DCLG website or alternatively, printed versions may be purchased from numerous sources.

Statutory Instrument 2005 No. 1541

The Regulatory Reform (Fire Safety) Order 2005

PART 1

GENERAL

Citation, commencement and extent
1.—(1) This Order may be cited as the Regulatory Reform (Fire Safety) Order 2005 and shall come into force in accordance with paragraphs (2) and (3).

(2) This article and article 52(1)(a) shall come into force on the day after the day on which this Order is made.

(3) The remaining provisions of this Order shall come into force on 1st April 2006.

(4) This Order extends to England and Wales only.

Interpretation
2. In this Order—

"alterations notice" has the meaning given by article 29;

"approved classification and labelling guide" means the Approved Guide to the Classification and Labelling of Dangerous Substances and Dangerous Preparations (5th edition) approved by the Health and Safety Commissionon 16th April 2002;

"the CHIP Regulations" means the Chemicals (Hazard Information and Packaging for Supply) Regulations 2002;

"child" means a person who is not over compulsory school age, construed in accordance with section 8 of the Education Act 1996;

"dangerous substance" means—

(a) a substance or preparation which meets the criteria in the approved classification and labelling guide for classification as a substance or preparation which is explosive, oxidising, extremely flammable, highly flammable or flammable, whether or not that substance or preparation is classified under the CHIP Regulations;

(b) a substance or preparation which because of its physico-chemical or chemical properties and the way it is used or is present in or on premises creates a risk; and

(c) any dust, whether in the form of solid particles or fibrous materials or otherwise, which can form an explosive mixture with air or an explosive atmosphere;

"domestic premises" means premises occupied as a private dwelling (including any garden, yard, garage, outhouse, or other appurtenance of such premises which is not used in common by the occupants of more than one such dwelling);

"employee" means a person who is or is treated as an employee for the purposes of the Health and Safety at Work etc. Act 1974 and related expressions are to be construed accordingly;

"enforcement notice" has the meaning given by article 30;

"enforcing authority" has the meaning given by article 25;

"explosive atmosphere" means a mixture, under atmospheric conditions, of air and one or more dangerous substances in the form of gases, vapours, mists or dusts in which, after ignition has occurred, combustion spreads to the entire unburned mixture;

"fire and rescue authority" means a fire and rescue authority under the Fire and Rescue Services Act 2004;

"fire inspector" means an inspector or assistant inspector appointed under section 28 of the Fire and Rescue Services Act 2004;

"general fire precautions" has the meaning given by article 4;

"hazard", in relation to a dangerous substance, means the physico-chemical or chemical property of that substance which has the potential to give rise to fire affecting the safety of a person, and references in this Order to "hazardous" are to be construed accordingly;

"inspector" means an inspector appointed under article 26 or a fire inspector;

"licensing authority" has the meaning given by article 42(3);

"normal ship-board activities" include the repair of a ship, save repair when carried out in dry dock;

"owner" means the person for the time being receiving the rackrent of the premises in connection with which the word is used, whether on his own account or as agent or trustee for another person, or who would so receive the rackrent if the premises were let at a rackrent;

"personal protective equipment" means all equipment which is intended to be worn or held by a person in or on premises and which protects that person against one or more risks to his safety, and any addition or accessory designed to meet that objective;

"place of safety" in relation to premises, means a safe area beyond the premises.

"premises" includes any place and, in particular, includes—

(a) any workplace;

(b) any vehicle, vessel, aircraft or hovercraft;

(c) any installation on land (including the foreshore and other land intermittently covered by water), and any other installation (whether floating, or resting on the seabed or the subsoil thereof, or resting on other land covered with water or the subsoil thereof); and

(d) any tent or movable structure;

"preparation" means a mixture or solution of two or more substances;

"preventive and protective measures" means the measures which have been identified by the responsible person in consequence of a risk assessment as the general fire precautions he needs to take to comply with the requirements and prohibitions imposed on him by or under this Order;

"prohibition notice" has the meaning given by article 31;

“public road” means a highway maintainable at public expense within the meaning of section 329 of the Highways Act 1980;

“rackrent” in relation to premises, means a rent that is not less than two-thirds of the rent at which the property might reasonably be expected to be let from year to year, free from all usual tenant’s rates and taxes, and deducting from it the probable average cost of the repairs, insurance and other expenses (if any) necessary to maintain the property in a state to command such rent;

“the relevant local authority”, in relation to premises, means—

(a) if the premises are in Greater London but are not in the City of London, the London Borough in the area of which the premises are situated;

(b) if the premises are in the City of London, the Common Council of the City of London;

(c) if the premises are in England in a metropolitan county, the district council in the area of which the premises are situated;

(d) if the premises are in England but are not in Greater London or a metropolitan county—

(i) the county council in the area of which the premises are situated; or

(ii) if there is no county council in the area of which the premises are situated, the district council in that area;

(e) if the premises are in Wales, the county council or county borough council in the area of which the premises are situated;

“relevant persons” means—

(a) any person (including the responsible person) who is or may be lawfully on the premises; and

(b) any person in the immediate vicinity of the premises who is at risk from a fire on the premises,

but does not include a fire-fighter who is carrying out his duties in relation to a function of a fire and rescue authority under section 7, 8 or 9 of the Fire and Rescue Services Act 2004 (fire-fighting, road traffic accidents and other emergencies), other than in relation to a function under section 7(2)(d), 8(2)(d) or 9(3)(d) of that Act;

“responsible person” has the meaning given by article 3;

“risk” means the risk to the safety of persons from fire;

“risk assessment” means the assessment required by article 9(1);

“safety” means the safety of persons in respect of harm caused by fire; and “safe” shall be interpreted accordingly;

“safety data sheet” means a safety data sheet within the meaning of regulation 5 of the CHIP Regulations;

“ship” includes every description of vessel used in navigation;

“special, technical and organisational measures” include—

(a) technical means of supervision;

(b) connecting devices;

(c) control and protection systems;

(d) engineering controls and solutions;

(e) equipment;

(f) materials;

(g) protective systems; and

(h) warning and other communication systems;

"substance" means any natural or artificial substance whether in solid or liquid form or in the form of a gas or vapour;

"visiting force" means any such body, contingent, or detachment of the forces of any country as is a visiting force for the purposes of any of the provisions of the Visiting Forces Act 1952;

"workplace" means any premises or parts of premises, not being domestic premises, used for the purposes of an employer's undertaking and which are made available to an employee of the employer as a place of work and includes—

(a) any place within the premises to which such employee has access while at work; and

(b) any room, lobby, corridor, staircase, road, or other place—

(i) used as a means of access to or egress from that place of work; or

(ii) where facilities are provided for use in connection with that place of work, other than a public road;

"young person" means any person who has not attained the age of 18.

Meaning of "responsible person"

3. In this Order "responsible person" means—

(a) in relation to a workplace, the employer, if the workplace is to any extent under his control;

(b) in relation to any premises not falling within paragraph (a)—

(i) the person who has control of the premises (as occupier or otherwise) in connection with the carrying on by him of a trade, business or other undertaking (for profit or not); or

(ii) the owner, where the person in control of the premises does not have control in connection with the carrying on by that person of a trade, business or other undertaking.

Meaning of "general fire precautions"

4. —(1) In this Order "general fire precautions" in relation to premises means, subject to paragraph (2)—

(a) measures to reduce the risk of fire on the premises and the risk of the spread of fire on the premises;

(b) measures in relation to the means of escape from the premises;

(c) measures for securing that, at all material times, the means of escape can be safely and effectively used;

(d) measures in relation to the means for fighting fires on the premises;

(e) measures in relation to the means for detecting fire on the premises and giving warning in case of fire on the premises; and

(f) measures in relation to the arrangements for action to be taken in the event of fire on the premises, including—

(i) measures relating to the instruction and training of employees; and

(ii) measures to mitigate the effects of the fire.

(2) The precautions referred to in paragraph (1) do not include special, technical or organisational measures required to be taken or observed in any workplace in connection with the carrying on of any work process, where those measures —

(a) are designed to prevent or reduce the likelihood of fire arising from such a work process or reduce its intensity; and

(b) are required to be taken or observed to ensure compliance with any requirement of the relevant statutory provisions within the meaning given by section 53(1) of the Health and Safety at Work etc 1974[12].

(3) In paragraph (2) "work process" means all aspects of work involving, or in connection with—

(a) the use of plant or machinery; or

(b) the use or storage of any dangerous substance.

Duties under this Order

5. —(1) Where the premises are a workplace, the responsible person must ensure that any duty imposed by articles 8 to 22 or by regulations made under article 24 is complied with in respect of those premises.

(2) Where the premises are not a workplace, the responsible person must ensure that any duty imposed by articles 8 to 22 or by regulations made under article 24 is complied with in respect of those premises, so far as the requirements relate to matters within his control.

(3) Any duty imposed by articles 8 to 22 or by regulations made under article 24 on the responsible person in respect of premises shall also be imposed on every person, other than the responsible person referred to in paragraphs (1) and (2), who has, to any extent, control of those premises so far as the requirements relate to matters within his control.

(4) Where a person has, by virtue of any contract or tenancy, an obligation of any extent in relation to—

(a) the maintenance or repair of any premises, including anything in or on premises; or

(b) the safety of any premises,

that person is to be treated, for the purposes of paragraph (3), as being a person who has control of the premises to the extent that his obligation so extends.

(5) Articles 8 to 22 and any regulations made under article 24 only require the taking or observance of general fire precautions in respect of relevant persons.

Application to premises

6. —(1) This Order does not apply in relation to —

(a) domestic premises, except to the extent mentioned in article 31(10);

(b) an offshore installation within the meaning of regulation 3 of the Offshore Installation and Pipeline Works (Management and Administration) Regulations 1995;

(c) a ship, in respect of the normal ship-board activities of a ship's crew which are carried out solely by the crew under the direction of the master;

(d) fields, woods or other land forming part of an agricultural or forestry undertaking but which is not inside a building and is situated away from the undertaking's main buildings;

(e) an aircraft, locomotive or rolling stock, trailer or semi-trailer used as a means of transport or a vehicle for which a licence is in force under the Vehicle Excise and Registration Act 1994 or a vehicle exempted from duty under that Act;

(f) a mine within the meaning of section 180 of the Mines and Quarries Act 1954, other than any building on the surface at a mine;

(g) a borehole site to which the Borehole Sites and Operations Regulations 1995 apply.

(2) Subject to the preceding paragraph of this article, this Order applies in relation to any premises.

Disapplication of certain provisions

7. —(1) Articles 9(4) and (5) and 19(2) do not apply in relation to occasional work or short-term work involving work regulated as not being harmful, damaging, or dangerous to young people in a family undertaking.

(2) Articles 9(2), 12, 16, 19(3) and 22(2) do not apply in relation to the use of means of transport by land, water or air where the use of means of transport is regulated by international agreements and the European Community directives giving effect to them and in so far as the use of means of transport falls within the disapplication in article 1.2(e) of Council Directive 1999/92/EC on minimum requirements for improving the safety and health of workers potentially at risk from explosive atmospheres, except for any means of transport intended for use in a potentially explosive atmosphere.

(3) Articles 19 and 21 impose duties only on responsible persons who are employers.

(4) The requirements of articles 8 to 23, or of any regulations made under article 24, do not have effect to the extent that they would prevent any of the following from carrying out their duties—

(a) any member of the armed forces of the Crown or of any visiting force;

(b) any constable or any member of a police force not being a constable;

(c) any member of any emergency service.

(5) Without prejudice to paragraph (4), article 14(2)(f) does not apply to any premises constituting, or forming part of, a prison within the meaning of the Prison Act 1952 or constituting, or forming part of, a remand centre, detention centre or youth custody centre provided by the Secretary of State under section 43 of that Act or any part of any other premises used for keeping persons in lawful custody or detention.

(6) Where paragraph (4) or (5) applies, the safety of relevant persons must nevertheless be ensured so far as is possible.

PART 2

FIRE SAFETY DUTIES

Duty to take general fire precautions

8. —(1) The responsible person must—

(a) take such general fire precautions as will ensure, so far as is reasonably practicable, the safety of any of his employees; and

(b) in relation to relevant persons who are not his employees, take such general fire precautions as may reasonably be required in the circumstances of the case to ensure that the premises are safe.

Risk assessment

9. —(1) The responsible person must make a suitable and sufficient assessment of the risks to which relevant persons are exposed for the purpose of identifying the general fire precautions he needs to take to comply with the requirements and prohibitions imposed on him by or under this Order.

(2) Where a dangerous substance is or is liable to be present in or on the premises, the risk assessment must include consideration of the matters set out in Part 1 of Schedule 1.

(3) Any such assessment must be reviewed by the responsible person regularly so as to keep it up to date and particularly if—

(a) there is reason to suspect that it is no longer valid; or

(b) there has been a significant change in the matters to which it relates including when the premises, special, technical and organisational measures, or organisation of the work undergo significant changes, extensions, or conversions,

and where changes to an assessment are required as a result of any such review, the responsible person must make them.

(4) The responsible person must not employ a young person unless he has, in relation to risks to young persons, made or reviewed an assessment in accordance with paragraphs (1) and (5).

(5) In making or reviewing the assessment, the responsible person who employs or is to employ a young person must take particular account of the matters set out in Part 2 of Schedule 1.

(6) As soon as practicable after the assessment is made or reviewed, the responsible person must record the information prescribed by paragraph (7) where—

(a) he employs five or more employees;

(b) a licence under an enactment is in force in relation to the premises; or

(c) an alterations notice requiring this is in force in relation to the premises.

(7) The prescribed information is—

(a) the significant findings of the assessment, including the measures which have been or will be taken by the responsible person pursuant to this Order; and

(b) any group of persons identified by the assessment as being especially at risk.

(8) No new work activity involving a dangerous substance may commence unless—

(a) the risk assessment has been made; and

(b) the measures required by or under this Order have been implemented.

Principles of prevention to be applied

10. Where the responsible person implements any preventive and protective measures he must do so on the basis of the principles specified in Part 3 of Schedule 1.

Fire safety arrangements

11. —(1) The responsible person must make and give effect to such arrangements as are appropriate, having regard to the size of his undertaking and the nature of its activities, for the effective planning, organisation, control, monitoring and review of the preventive and protective measures.

(2) The responsible person must record the arrangements referred to in paragraph (1) where—

(a) he employs five or more employees;

(b) a licence under an enactment is in force in relation to the premises; or

(c) an alterations notice requiring a record to be made of those arrangements is in force in relation to the premises.

Elimination or reduction of risks from dangerous substances

12. —(1) Where a dangerous substance is present in or on the premises, the responsible person must ensure that risk to relevant persons related to the presence of the substance is either eliminated or reduced so far as is reasonably practicable.

(2) In complying with his duty under paragraph (1), the responsible person must, so far as is reasonably practicable, replace a dangerous substance, or the use of a dangerous substance, with a substance or process which either eliminates or reduces the risk to relevant persons.

(3) Where it is not reasonably practicable to eliminate risk pursuant to paragraphs (1) and (2), the responsible person must, so far as is reasonably practicable, apply measures consistent with the risk assessment and appropriate to the nature of the activity or operation, including the measures specified in Part 4 of Schedule 1 to this Order to—

(a) control the risk, and

(b) mitigate the detrimental effects of a fire.

(4) The responsible person must—

(a) arrange for the safe handling, storage and transport of dangerous substances and waste containing dangerous substances; and

(b) ensure that any conditions necessary pursuant to this Order for ensuring the elimination or reduction of risk are maintained.

Fire-fighting and fire detection

13. —(1) Where necessary (whether due to the features of the premises, the activity carried on there, any hazard present or any other relevant circumstances) in order to safeguard the safety of relevant persons, the responsible person must ensure that—

(a) the premises are, to the extent that it is appropriate, equipped with appropriate fire-fighting equipment and with fire detectors and alarms; and

(b) any non-automatic fire-fighting equipment so provided is easily accessible, simple to use and indicated by signs.

(2) For the purposes of paragraph (1) what is appropriate is to be determined having regard to the dimensions and use of the premises, the equipment contained on the premises, the physical and chemical properties of the substances likely to be present and the maximum number of persons who may be present at any one time.

(3) The responsible person must, where necessary—

(a) take measures for fire-fighting in the premises, adapted to the nature of the activities carried on there and the size of the undertaking and of the premises concerned;

(b) nominate competent persons to implement those measures and ensure that the number of such persons, their training and the equipment available to them are adequate, taking into account the size of, and the specific hazards involved in, the premises concerned; and

(c) arrange any necessary contacts with external emergency services, particularly as regards fire-fighting, rescue work, first-aid and emergency medical care.

(4) A person is to be regarded as competent for the purposes of paragraph (3)(b) where he has sufficient training and experience or knowledge and other qualities to enable him properly to implement the measures referred to in that paragraph.

Emergency routes and exits

14. —(1) Where necessary in order to safeguard the safety of relevant persons, the responsible person must ensure that routes to emergency exits from premises and the exits themselves are kept clear at all times.

(2) The following requirements must be complied with in respect of premises where necessary (whether due to the features of the premises, the activity carried on there, any hazard present or any other relevant circumstances) in order to safeguard the safety of relevant persons—

(a) emergency routes and exits must lead as directly as possible to a place of safety;

(b) in the event of danger, it must be possible for persons to evacuate the premises as quickly and as safely as possible;

(c) the number, distribution and dimensions of emergency routes and exits must be adequate having regard to the use, equipment and dimensions of the premises and the maximum number of persons who may be present there at any one time;

(d) emergency doors must open in the direction of escape;

(e) sliding or revolving doors must not be used for exits specifically intended as emergency exits;

(f) emergency doors must not be so locked or fastened that they cannot be easily and immediately opened by any person who may require to use them in an emergency;

(g) emergency routes and exits must be indicated by signs; and

(h) emergency routes and exits requiring illumination must be provided with emergency lighting of adequate intensity in the case of failure of their normal lighting.

Procedures for serious and imminent danger and for danger areas

15. —(1) The responsible person must—

(a) establish and, where necessary, give effect to appropriate procedures, including safety drills, to be followed in the event of serious and imminent danger to relevant persons;

(b) nominate a sufficient number of competent persons to implement those procedures in so far as they relate to the evacuation of relevant persons from the premises; and

(c) ensure that no relevant person has access to any area to which it is necessary to restrict access on grounds of safety, unless the person concerned has received adequate safety instruction.

(2) Without prejudice to the generality of paragraph (1)(a), the procedures referred to in that sub-paragraph must—

(a) so far as is practicable, require any relevant persons who are exposed to serious and imminent danger to be informed of the nature of the hazard and of the steps taken or to be taken to protect them from it;

(b) enable the persons concerned (if necessary by taking appropriate steps in the absence of guidance or instruction and in the light of their knowledge and the technical means at their disposal) to stop work and immediately proceed to a place of safety in the event of their being exposed to serious, imminent and unavoidable danger; and

(c) save in exceptional cases for reasons duly substantiated (which cases and reasons must be specified in those procedures), require the persons concerned to be prevented from resuming work in any situation where there is still a serious and imminent danger.

(3) A person is to be regarded as competent for the purposes of paragraph (1) where he has sufficient training and experience or knowledge and other qualities to enable him properly to implement the evacuation procedures referred to in that paragraph.

Additional emergency measures in respect of dangerous substances

16. —(1) Subject to paragraph (4), in order to safeguard the safety of relevant persons arising from an accident, incident or emergency related to the presence of a dangerous substance in or on the premises, the responsible person must ensure that—

(a) information on emergency arrangements is available, including—

(i) details of relevant work hazards and hazard identification arrangements; and

(ii) specific hazards likely to arise at the time of an accident, incident or emergency;

(b) suitable warning and other communication systems are established to enable an appropriate response, including remedial actions and rescue operations, to be made immediately when such an event occurs;

(c) where necessary, before any explosion conditions are reached, visual or audible warnings are given and relevant persons withdrawn; and

(d) where the risk assessment indicates it is necessary, escape facilities are provided and maintained to ensure that, in the event of danger, relevant persons can leave endangered places promptly and safely.

(2) Subject to paragraph (4), the responsible person must ensure that the information required by article 15(1)(a) and paragraph (1)(a) of this article, together with information on the matters referred to in paragraph (1)(b) and (d) is—

(a) made available to relevant accident and emergency services to enable those services, whether internal or external to the premises, to prepare their own response procedures and precautionary measures; and

(b) displayed at the premises, unless the results of the risk assessment make this unnecessary.

(3) Subject to paragraph (4), in the event of a fire arising from an accident, incident or emergency related to the presence of a dangerous substance in or on the premises, the responsible person must ensure that—

(a) immediate steps are taken to—

(i) mitigate the effects of the fire;

(ii) restore the situation to normal; and

(iii) inform those relevant persons who may be affected; and

(b) only those persons who are essential for the carrying out of repairs and other necessary work are permitted in the affected area and they are provided with—

(i) appropriate personal protective equipment and protective clothing; and

(ii) any necessary specialised safety equipment and plant,

which must be used until the situation is restored to normal.

(4) Paragraphs (1) to (3) do not apply where—

(a) the results of the risk assessment show that, because of the quantity of each dangerous substance in or on the premises, there is only a slight risk to relevant persons; and

(b) the measures taken by the responsible person to comply with his duty under article 12 are sufficient to control that risk.

Maintenance

17. —(1) Where necessary in order to safeguard the safety of relevant persons the responsible person must ensure that the premises and any facilities, equipment and devices provided in respect of the premises under this Order or, subject to paragraph (6), under any other enactment, including any enactment repealed or revoked by this Order, are subject to a suitable system of maintenance and are maintained in an efficient state, in efficient working order and in good repair.

(2) Where the premises form part of a building, the responsible person may make arrangements with the occupier of any other premises forming part of the building for the purpose of ensuring that the requirements of paragraph (1) are met.

(3) Paragraph (2) applies even if the other premises are not premises to which this Order applies.

(4) The occupier of the other premises must co-operate with the responsible person for the purposes of paragraph (2).

(5) Where the occupier of the other premises is not also the owner of those premises, the references to the occupier in paragraphs (2) and (4) are to be taken to be references to both the occupier and the owner.

(6) Paragraph (1) only applies to facilities, equipment and devices provided under other enactments where they are provided in connection with general fire precautions.

Safety assistance

18. —(1) The responsible person must, subject to paragraphs (6) and (7), appoint one or more competent persons to assist him in undertaking the preventive and protective measures.

(2) Where the responsible person appoints persons in accordance with paragraph (1), he must make arrangements for ensuring adequate co-operation between them.

(3) The responsible person must ensure that the number of persons appointed under paragraph (1), the time available for them to fulfil their functions and the means at their disposal are adequate having regard to the size of the premises, the risks to which relevant persons are exposed and the distribution of those risks throughout the premises.

(4) The responsible person must ensure that—

(a) any person appointed by him in accordance with paragraph (1) who is not in his employment—

(i) is informed of the factors known by him to affect, or suspected by him of affecting, the safety of any other person who may be affected by the conduct of his undertaking; and

(ii) has access to the information referred to in article 19(3); and

(b) any person appointed by him in accordance with paragraph (1) is given such information about any person working in his undertaking who is—

(i) employed by him under a fixed-term contract of employment, or

(ii) employed in an employment business,

as is necessary to enable that person properly to carry out the function specified in that paragraph.

(5) A person is to be regarded as competent for the purposes of this article where he has sufficient training and experience or knowledge and other qualities to enable him properly to assist in undertaking the preventive and protective measures.

(6) Paragraph (1) does not apply to a self-employed employer who is not in partnership with any other person, where he has sufficient training and experience or knowledge and other qualities properly to assist in undertaking the preventive and protective measures.

(7) Paragraph (1) does not apply to individuals who are employers and who are together carrying on business in partnership, where at least one of the individuals concerned has sufficient training and experience or knowledge and other qualities—

(a) properly to undertake the preventive and protective measures; and

(b) properly to assist his fellow partners in undertaking those measures.

(8) Where there is a competent person in the responsible person's employment, that person must be appointed for the purposes of paragraph (1) in preference to a competent person not in his employment.

Provision of information to employees

19. —(1) The responsible person must provide his employees with comprehensible and relevant information on—

(a) the risks to them identified by the risk assessment;

(b) the preventive and protective measures;

(c) the procedures and the measures referred to in article 15(1)(a);

(d) the identities of those persons nominated by him in accordance with article 13(3)(b) or appointed in accordance with article 15(1)(b) ; and

(e) the risks notified to him in accordance with article 22(1)(c).

(2) The responsible person must, before employing a child, provide a parent of the child with comprehensible and relevant information on—

(a) the risks to that child identified by the risk assessment;

(b) the preventive and protective measures; and

(c) the risks notified to him in accordance with article 22(1)(c),

and for the purposes of this paragraph, "parent of the child" includes a person who has parental responsibility, within the meaning of section 3 of the Children Act 1989, for the child.

(3) Where a dangerous substance is present in or on the premises, the responsible person must, in addition to the information provided under paragraph (1) provide his employees with —

(a) the details of any such substance including—

(i) the name of the substance and the risk which it presents;

(ii) access to any relevant safety data sheet; and

(iii) legislative provisions (concerning the hazardous properties of any such substance) which apply to the substance; and

(b) the significant findings of the risk assessment.

(4) The information required by paragraph (3) must be—

(a) adapted to take account of significant changes in the activity carried out or methods or work used by the responsible person; and

(b) provided in a manner appropriate to the risk identified by the risk assessment.

Provision of information to employers and the self-employed from outside undertakings

20. —(1) The responsible person must ensure that the employer of any employees from an outside undertaking who are working in or on the premises is provided with comprehensible and relevant information on—

(a) the risks to those employees; and

(b) the preventive and protective measures taken by the responsible person.

(2) The responsible person must ensure that any person working in his undertaking who is not his employee is provided with appropriate instructions and comprehensible and relevant information regarding any risks to that person.

(3) The responsible person must—

(a) ensure that the employer of any employees from an outside undertaking who are working in or on the premises is provided with sufficient information to enable that employer to identify any person nominated by the responsible person in accordance with article 15 (1)(b) to implement evacuation procedures as far as those employees are concerned; and

(b) take all reasonable steps to ensure that any person from an outside undertaking who is working in or on the premises receives sufficient information to enable that person to identify any person nominated by the responsible person in accordance with article 15 (1)(b) to implement evacuation procedures as far as they are concerned.

Training

21. —(1) The responsible person must ensure that his employees are provided with adequate safety training—

(a) at the time when they are first employed; and

(b) on their being exposed to new or increased risks because of—

(i) their being transferred or given a change of responsibilities within the responsible person's undertaking;

(ii) the introduction of new work equipment into, or a change respecting work equipment already in use within, the responsible person's undertaking;

(iii) the introduction of new technology into the responsible person's undertaking; or

(iv) the introduction of a new system of work into, or a change respecting a system of work already in use within, the responsible person's undertaking.

(2) The training referred to in paragraph (1) must—

(a) include suitable and sufficient instruction and training on the appropriate precautions and actions to be taken by the employee in order to safeguard himself and other relevant persons on the premises;

(b) be repeated periodically where appropriate;

(c) be adapted to take account of any new or changed risks to the safety of the employees concerned;

(d) be provided in a manner appropriate to the risk identified by the risk assessment; and

(e) take place during working hours.

Co-operation and co-ordination

22. —(1) Where two or more responsible persons share, or have duties in respect of, premises (whether on a temporary or a permanent basis) each such person must—

(a) co-operate with the other responsible person concerned so far as is necessary to enable them to comply with the requirements and prohibitions imposed on them by or under this Order;

(b) (taking into account the nature of his activities) take all reasonable steps to co-ordinate the measures he takes to comply with the requirements and prohibitions imposed on him by or under this Order with the measures the other responsible persons are taking to comply with the requirements and prohibitions imposed on them by or under this Order; and

(c) take all reasonable steps to inform the other responsible persons concerned of the risks to relevant persons arising out of or in connection with the conduct by him of his undertaking.

(2) Where two or more responsible persons share premises (whether on a temporary or a permanent basis) where an explosive atmosphere may occur, the responsible person who has overall responsibility for the premises must co-ordinate the implementation of all the measures required by this Part to be taken to protect relevant persons from any risk from the explosive atmosphere.

General duties of employees at work

23. —(1) Every employee must, while at work—

(a) take reasonable care for the safety of himself and of other relevant persons who may be affected by his acts or omissions at work;

(b) as regards any duty or requirement imposed on his employer by or under any provision of this Order, co-operate with him so far as is necessary to enable that duty or requirement to be performed or complied with; and

(c) inform his employer or any other employee with specific responsibility for the safety of his fellow employees—

(i) of any work situation which a person with the first-mentioned employee's training and instruction would reasonably consider represented a serious and immediate danger to safety; and

(ii) of any matter which a person with the first-mentioned employee's training and instruction would reasonably consider represented a shortcoming in the employer's protection arrangements for safety,

in so far as that situation or matter either affects the safety of that first-mentioned employee or arises out of or in connection with his own activities at work, and has not previously been reported to his employer or to any other employee of that employer in accordance with this sub-paragraph.

Power to make regulations about fire precautions

24. —(1) The Secretary of State may by regulations make provision as to the precautions which are to be taken or observed in relation to the risk to relevant persons as regards premises in relation to which this Order applies.

(2) Without prejudice to the generality of paragraph (1), regulations made by the Secretary of State may impose requirements—

(a) as to the provision, maintenance and keeping free from obstruction of any means of escape in case of fire;

(b) as to the provision and maintenance of means for securing that any means of escape can be safely and effectively used at all material times;

(c) as to the provision and maintenance of means for fighting fire and means for giving warning in case of fire;

(d) as to the internal construction of the premises and the materials used in that construction;

(e) for prohibiting altogether the presence or use in the premises of furniture or equipment of any specified description, or prohibiting its presence or use unless specified standards or conditions are complied with;

(f) for securing that persons employed to work in the premises receive appropriate instruction or training in what to do in case of fire;

(g) for securing that, in specified circumstances, specified numbers of attendants are stationed in specified parts of the premises; and

(h) as to the keeping of records of instruction or training given, or other things done, in pursuance of the regulations.

(3) Regulations under this article—

(a) may impose requirements on persons other than the responsible person; and

(b) may, as regards any of their provisions, make provision as to the person or persons who is or are to be responsible for any contravention of that provision.

(4) The Secretary of State must, before making any regulations under this article, consult with such persons or bodies of persons as appear to him to be appropriate.

(5) The power of the Secretary of State to make regulations under this article—

(a) is exercisable by statutory instrument, which is subject to annulment in pursuance of a resolution of either House of Parliament;

(b) includes power to make different provision in relation to different circumstances; and

(c) includes power to grant or provide for the granting of exemptions from any of the provisions of the regulations, either unconditionally or subject to conditions.

PART 3

ENFORCEMENT

Enforcing authorities

25. For the purposes of this Order, "enforcing authority" means—

(a) the fire and rescue authority for the area in which premises are, or are to be, situated, in any case not falling within any of sub-paragraphs (b) to (e);

(b) the Health and Safety Executive in relation to—

(i) any premises for which a licence is required in accordance with section 1 of the Nuclear Installations Act 1965 or for which a permit is required in accordance with section 2 of that Act;

(ii) any premises which would, except for the fact that it is used by, or on behalf of, the Crown, be required to have a licence or permit in accordance with the provisions referred to in subparagraph (i);

(iii) a ship, including a ship belonging to Her Majesty which forms part of Her Majesty's Navy, which is in the course of construction, reconstruction or conversion or repair by persons who include persons other than the master and crew of the ship;

(iv) any workplace which is or is on a construction site within the meaning of regulation 2(1) of the Construction (Health, Safety and Welfare) Regulations 1996 and to which those Regulations apply, other than construction sites referred to in regulation 33 of those Regulations.

(c) the fire service maintained by the Secretary of State for Defence in relation to—

(i) premises, other than premises falling within paragraph (b)(iii), occupied solely for the purposes of the armed forces of the Crown;

(ii) premises occupied solely by any visiting force or an international headquarters or defence organisation designated for the purposes of the International Headquarters and Defence Organisations Act 1964;

(iii) premises, other than premises falling within paragraph (b)(iii), which are situated within premises occupied solely for the purposes of the armed forces of the Crown but which are not themselves so occupied;

(d) the relevant local authority in relation to premises which consist of—

(i) a sports ground designated as requiring a safety certificate under section 1 of the Safety of Sports Grounds Act 1975 (safety certificates for large sports stadia);

(ii) a regulated stand within the meaning of section 26(5) of the Fire Safety and Safety of Places of Sport Act 1987 (safety certificates for stands at sports grounds);

(e) a fire inspector, or any person authorised by the Secretary of State to act for the purposes of this Order, in relation to—

(i) premises owned or occupied by the Crown, other than premises falling within paragraph (b)(ii) and (c));

(ii) premises in relation to which the United Kingdom Atomic Energy Authority is the responsible person, other than premises falling within paragraph (b)(ii)).

Enforcement of Order

26. —(1) Every enforcing authority must enforce the provisions of this Order and any regulations made under it in relation to premises for which it is the enforcing authority and for that purpose, except where a fire inspector or other person authorised by the Secretary of State is the enforcing authority, may appoint inspectors.

(2) In performing the duty imposed by paragraph (1), the enforcing authority must have regard to such guidance as the Secretary of State may give it.

(3) A fire and rescue authority has power to arrange with the Health and Safety Commission or the Office of Rail Regulation for such of the authority's functions under this Order as may be specified in the arrangements to be performed on its behalf by the Health and Safety Executive or the Office of Rail Regulation, as the case may be, (with or without payment) in relation to any particular workplace.

Powers of inspectors

27. —(1) Subject to the provisions of this article, an inspector may do anything necessary for the purpose of carrying out this Order and any regulations made under it into effect and in particular, so far as may be necessary for that purpose, shall have power to do at any reasonable time the following—

(a) to enter any premises which he has reason to believe it is necessary for him to enter for the purpose mentioned above and to inspect the whole or part of the premises and anything in them, where such entry and inspection may be effected without the use of force;

(b) to make such inquiry as may be necessary for any of the following purposes—

(i) to ascertain, as regards any premises, whether the provisions of this Order or any regulations made under it apply or have been complied with; and

(ii) to identify the responsible person in relation to the premises;

(c) to require the production of, or where the information is recorded in computerised form, the furnishing of extracts from, any records (including plans)—

(i) which are required to be kept by virtue of any provision of this Order or regulations made under it; or

(ii) which it is necessary for him to see for the purposes of an examination or inspection under this article,

and to inspect and take copies of, or of any entry in, the records;

(d) to require any person having responsibilities in relation to any premises (whether or not the responsible person) to give him such facilities and assistance with respect to any matters or things to which the responsibilities of that person extend as are necessary for the purpose of enabling the inspector to exercise any of the powers conferred on him by this article;

(e) to take samples of any articles or substances found in any premises which he has power to enter for the purpose of ascertaining their fire resistance or flammability; and

(f) in the case of any article or substance found in any premises which he has power to enter, being an article or substance which appears to him to have caused or to be likely to cause danger to the safety of relevant persons, to cause it to be dismantled or subjected to any process or test (but not so as to damage or destroy it unless this is, in the circumstances, necessary).

(2) An inspector must, if so required when visiting any premises in the exercise of powers conferred by this article, produce to the occupier of the premises evidence of his authority.

(3) Where an inspector proposes to exercise the power conferred by paragraph (1)(f) he must, if requested by a person who at the time is present in and has responsibilities in relation to those premises, cause anything which is to be done by virtue of that power to be done in the presence of that person.

(4) Before exercising the power conferred by paragraph (1)(f) an inspector must consult such persons as appear to him appropriate for the purpose of ascertaining what dangers, if any, there may be in doing anything which he proposes to do under that power.

Exercise on behalf of fire inspectors etc. of their powers by officers of fire brigades

28. —(1) The powers conferred by article 27 on a fire inspector, or any other person authorised by the Secretary of State under article 25(e), are also exercisable by an employee of the fire and rescue authority when authorised in writing by such an inspector for the purpose of reporting to him on any matter falling within his functions under this Order; and articles 27(2) and (3) and 32(2)(d) to (f), with the necessary modifications, apply accordingly.

(2) A fire inspector, or other person authorised by the Secretary of State, must not authorise an employee of a fire and rescue authority under this article except with the consent of the fire and rescue authority.

Alterations notices

29. —(1) The enforcing authority may serve on the responsible person a notice (in this Order referred to as "an alterations notice") if the authority is of the opinion that the premises—

(a) constitute a serious risk to relevant persons (whether due to the features of the premises, their use, any hazard present, or any other circumstances); or

(b) may constitute such a risk if a change is made to them or the use to which they are put.

(2) An alterations notice must—

(a) state that the enforcing authority is of the opinion referred to in paragraph (1); and

(b) specify the matters which in their opinion, constitute a risk to relevant persons or may constitute such a risk if a change is made to the premises or the use to which they are put.

(3) Where an alterations notice has been served in respect of premises, the responsible person must, before making any of the changes specified in paragraph (4) which may result in a significant increase in risk, notify the enforcing authority of the proposed changes.

(4) The changes referred to in paragraph (3) are—

(a) a change to the premises;

(b) a change to the services, fittings or equipment in or on the premises;

(c) an increase in the quantities of dangerous substances which are present in or on the premises;

(d) a change to the use of the premises.

(5) An alterations notice may include a requirement that, in addition to the notification required by paragraph (3), the responsible person must —

(a) take all reasonable steps to notify the terms of the notice to any other person who has duties under article 5(3) in respect of the premises;

(b) record the information prescribed in article 9(7), in accordance with article 9(6);

(c) record the arrangements required by article 11(1), in accordance with article 11(2); and

(d) before making the changes referred to in paragraph (3), send the enforcing authority the following —

(i) a copy of the risk assessment; and

(ii) a summary of the changes he proposes to make to the existing general fire precautions.

(6) An alterations notice served under paragraph (1) may be withdrawn at any time and, for the purposes of this article, the notice is deemed to be in force until such time as it is withdrawn or cancelled by the court under article 35(2).

(7) Nothing in this article prevents an enforcing authority from serving an enforcement notice or a prohibition notice in respect of the premises.

Enforcement notices

30. —(1) If the enforcing authority is of the opinion that the responsible person or any other person mentioned in article 5(3) has failed to comply with any provision of this Order or of any regulations made under it, the authority may, subject to article 36, serve on that person a notice (in this Order referred to as "an enforcement notice").

(2) An enforcement notice must—

(a) state that the enforcing authority is of the opinion referred to in paragraph (1) and why;

(b) specify the provisions which have not been complied with; and

(c) require that person to take steps to remedy the failure within such period from the date of service of the notice (not being less than 28 days) as may be specified in the notice.

(3) An enforcement notice may, subject to article 36, include directions as to the measures which the enforcing authority consider are necessary to remedy the failure referred to in paragraph (1) and any such measures may be framed so as to afford the person on whom the notice is served a choice between different ways of remedying the contravention.

(4) Where the enforcing authority is of the opinion that a person's failure to comply with this Order also extends to a workplace, or employees who work in a workplace, to which this Order applies but for which they are not the enforcing authority, the notice served by them under paragraph (1) may include requirements concerning that workplace or those employees; but before including any such requirements the enforcing authority must consult the enforcing authority for that workplace.

(5) Before serving an enforcement notice which would oblige a person to make an alteration to premises, the enforcing authority must consult—

(a) in cases where the relevant local authority is not the enforcing authority, the relevant local authority;

(b) in the case of premises used as a workplace which are within the field of responsibility of one or more enforcing authorities within the meaning of Part 1 of the Health and Safety at Work etc Act 1974, that authority or those authorities; and section 18(7) of the Health and Safety at Work etc Act 1974 (meaning in Part I of that Act of "enforcing authority" and of such an authority's "field of responsibility") applies for the purposes of this article as it applies for the purposes of that Part;

(c) in the case of a building or structure in relation to all or any part of which an initial notice given under section 47 of the Building Act 1984 is in force, the approved inspector who gave that initial notice;

(d) in the case of premises which are, include, or form part of, a designated sports ground or a sports ground at which there is a regulated stand, the relevant local authority, where that authority is not the enforcing authority; and for the purposes of this sub-paragraph, "sports ground" and "designated sports ground" have the same meaning as in the Safety of Sports Grounds Act 1975 and "regulated stand" has the same meaning as in the Fire Safety and Safety of Places of Sport Act 1987;

(e) any other person whose consent to the alteration would be required by or under any enactment.

(6) Without prejudice to the power of the court to cancel or modify an enforcement notice under article 35(2), no failure on the part of an enforcing authority to consult under paragraphs (4) or (5) makes an enforcement notice void.

(7) Where an enforcement notice has been served under paragraph (1)—

(a) the enforcing authority may withdraw the notice at any time before the end of the period specified in the notice; and

(b) if an appeal against the notice is not pending, the enforcing authority may extend or further extend the period specified in the notice.

Prohibition notices

31. —(1) If the enforcing authority is of the opinion that use of premises involves or will involve a risk to relevant persons so serious that use of the premises ought to be prohibited or restricted, the authority may serve on the responsible person or any other person mentioned in article 5(3) a notice (in this Order referred to as "a prohibition notice").

(2) The matters relevant to the assessment by the enforcing authority, for the purposes of paragraph (1), of the risk to relevant persons include anything affecting their escape from the premises in the event of fire.

(3) A prohibition notice must—

(a) state that the enforcing authority is of the opinion referred to in paragraph (1);

(b) specify the matters which in their opinion give or, as the case may be, will give rise to that risk; and

(c) direct that the use to which the prohibition notice relates is prohibited or restricted to such extent as may be specified in the notice until the specified matters have been remedied.

(4) A prohibition notice may include directions as to the measures which will have to be taken to remedy the matters specified in the notice and any such measures may be framed so as to afford the person on whom the notice is served a choice between different ways of remedying the matters.

(5) A prohibition or restriction contained in a prohibition notice pursuant to paragraph (3)(c) takes effect immediately it is served if the enforcing authority is of the opinion, and so states in the notice, that the risk of serious personal injury is or, as the case may be, will be imminent, and in any other case takes effect at the end of the period specified in the prohibition notice.

(6) Before serving a prohibition notice in relation to a house in multiple occupation, the enforcing authority shall, where practicable, notify the local housing authority of their intention and the use which they intend to prohibit or restrict.

(7) For the purposes of paragraph (6)—

"house in multiple occupation" means a house in multiple occupation as defined by sections 254 to 259 of the Housing Act 2004, as they have effect for the purposes of Part 1 of that Act (that is, without the exclusions contained in Schedule 14 to that Act);

and "local housing authority" has the same meaning as in section 261(2) of the Housing Act 2004.

(8) Without prejudice to the power of the court to cancel or modify a prohibition notice under article 35(2), no failure on the part of an enforcing authority to notify under paragraph (6) makes a prohibition notice void.

(9) Where a prohibition notice has been served under paragraph (1) the enforcing authority may withdraw it at any time.

(10) In this article, "premises" includes domestic premises other than premises consisting of or comprised in a house which is occupied as a single private dwelling and article 27 (powers of inspectors) shall be construed accordingly.

PART 4

OFFENCES AND APPEALS

Offences

32. —(1) It is an offence for any responsible person or any other person mentioned in article 5(3) to—

(a) fail to comply with any requirement or prohibition imposed by articles 8 to 22 and 38 (fire safety duties) where that failure places one or more relevant persons at risk of death or serious injury in case of fire;

(b) fail to comply with any requirement or prohibition imposed by regulations made, or having effect as if made, under article 24 where that failure places one or more relevant persons at risk of death or serious injury in case of fire;

(c) fail to comply with any requirement imposed by article 29(3) or (4) (alterations notices);

(d) fail to comply with any requirement imposed by an enforcement notice;

(e) fail, without reasonable excuse, in relation to apparatus to which article 37 applies (luminous tube signs)—

(i) to ensure that such apparatus which is installed in premises complies with article 37 (3) and (4);

(ii) to give a notice required by article 37(6) or (8), unless he establishes that some other person duly gave the notice in question;

(iii) to comply with a notice served under article 37(9).

(2) It is an offence for any person to—

(a) fail to comply with article 23 (general duties of employees at work) where that failure places one or more relevant persons at risk of death or serious injury in case of fire;

(b) make in any register, book, notice or other document required to be kept, served or given by or under, this Order, an entry which he knows to be false in a material particular;

(c) give any information which he knows to be false in a material particular or recklessly give any information which is so false, in purported compliance with any obligation to give information to which he is subject under or by virtue of this Order, or in response to any inquiry made by virtue of article 27(1)(b);

(d) obstruct, intentionally, an inspector in the exercise or performance of his powers or duties under this Order;

(e) fail, without reasonable excuse, to comply with any requirements imposed by an inspector under article 27(1)(c) or (d);

(f) pretend, with intent to deceive, to be an inspector;

(g) fail to comply with the prohibition imposed by article 40 (duty not to charge employees);

(h) fail to comply with any prohibition or restriction imposed by a prohibition notice.

(3) Any person guilty of an offence under paragraph (1)(a) to (d) and (2)(h) is liable—

(a) on summary conviction to a fine not exceeding the statutory maximum; or

(b) on conviction on indictment, to a fine, or to imprisonment for a term not exceeding two years, or to both.

(4) Any person guilty of an offence under paragraph (1)(e)(i) to (iii) is liable on summary conviction to a fine not exceeding level 3 on the standard scale.

(5) Any person guilty of an offence under paragraph (2)(a) is liable—

(a) on summary conviction to a fine not exceeding the statutory maximum; or

(b) on conviction on indictment, to a fine.

(6) Any person guilty of an offence under paragraph (2)(b), (c), (d) or (g) is liable on summary conviction to a fine not exceeding level 5 on the standard scale.

(7) Any person guilty of an offence under paragraph (2)(e) or (f) is liable on summary conviction to a fine not exceeding level 3 on the standard scale.

(8) Where an offence under this Order committed by a body corporate is proved to have been committed with the consent or connivance of, or to be attributable to any neglect on the part of, any director, manager,

secretary or other similar officer of the body corporate, or any person purporting to act in any such capacity, he as well as the body corporate is guilty of that offence, and is liable to be proceeded against and punished accordingly.

(9) Where the affairs of a body corporate are managed by its members, paragraph (8) applies in relation to the acts and defaults of a member in connection with his functions of management as if he were a director of the body corporate.

(10) Where the commission by any person of an offence under this Order, is due to the act or default of some other person, that other person is guilty of the offence, and a person may be charged with and convicted of the offence by virtue of this paragraph whether or not proceedings are taken against the first-mentioned person.

(11) Nothing in this Order operates so as to afford an employer a defence in any criminal proceedings for a contravention of those provisions by reason of any act or default of—

(a) an employee of his; or

(b) a person nominated under articles 13(3)(b) or 15(1)(b) or appointed under 18(1).

Defence

33. Subject to article 32(11), in any proceedings for an offence under this Order, except for a failure to comply with articles 8(1)(a) or 12, it is a defence for the person charged to prove that he took all reasonable precautions and exercised all due diligence to avoid the commission of such an offence.

Onus of proving limits of what is practicable or reasonably practicable

34. In any proceedings for an offence under this Order consisting of a failure to comply with a duty or requirement so far as is practicable or so far as is reasonably practicable, it is for the accused to prove that it was not practicable or reasonably practicable to do more than was in fact done to satisfy the duty or requirement.

Appeals

35. —(1) A person on whom an alterations notice, an enforcement notice, a prohibition notice or a notice given by the fire and rescue authority under article 37 (fire-fighters' switches for luminous tube signs) is served may, within 21 days from the day on which the notice is served, appeal to the court.

(2) On an appeal under this article the court may either cancel or affirm the notice, and if it affirms it, may do so either in its original form or with such modifications as the court may in the circumstances think fit.

(3) Where an appeal is brought against an alterations notice or an enforcement notice, the bringing of the appeal has the effect of suspending the operation of the notice until the appeal is finally disposed of or, if the appeal is withdrawn, until the withdrawal of the appeal.

(4) Where an appeal is brought against a prohibition notice, the bringing of the appeal does not have the effect of suspending the operation of the notice, unless, on the application of the appellant, the court so directs (and then only from the giving of the direction).

(5) In this article "the court" means a magistrates' court.

(6) The procedure for an appeal under paragraph (1) is by way of complaint for an order, and—

(a) the Magistrates' Courts Act 1980 applies to the proceedings; and

(b) the making of the complaint is deemed to be the bringing of the appeal.

(7) A person aggrieved by an order made by a magistrates' court on determining a complaint under this Order may appeal to the Crown Court; and for the avoidance of doubt, an enforcing authority may be a person aggrieved for the purposes of this paragraph.

Determination of disputes by Secretary of State

36. —(1) This article applies where—

(a) a responsible person or any other person mentioned in article 5(3) has failed to comply with any provision of this Order or of any regulations made under it; and

(b) the enforcing authority and that person cannot agree on the measures which are necessary to remedy the failure.

(2) Where this article applies, the enforcing authority and the person referred to in paragraph (1)(a) may agree to refer the question as to what measures are necessary to remedy the failure referred to in paragraph (1)(a) to the Secretary of State for his determination.

(3) The Secretary of State may, by notice in writing to both parties, require the provision of such further information, including plans, specified in the notice, within the period so specified, as the Secretary of State may require for the purpose of making a determination.

(4) If the information required under paragraph (3) is not provided within the period specified, the Secretary of State may refuse to proceed with the determination.

(5) Where the Secretary of State has made a determination under this article, the enforcing authority may not, subject to paragraph (6), take any enforcement action the effect of which would be to conflict with his determination; and in this article, "enforcement action" means the service of an enforcement notice or the inclusion of any directions in an enforcement notice.

(6) Paragraph (5) does not apply where, since the date of the determination by the Secretary of State, there has been a change to the premises or the use to which they are put such that the risk to relevant persons has significantly changed.

PART 5

MISCELLANEOUS

Fire-fighters' switches for luminous tube signs etc.

37. —(1) Subject to paragraph (11), this article applies to apparatus consisting of luminous tube signs designed to work at a voltage normally exceeding the prescribed voltage, or other equipment so designed, and references in this article to a cut-off switch are, in a case where a transformer is provided to raise the voltage to operate the apparatus, references to a cut-off switch on the low-voltage side of the transformer.

(2) In paragraph (1) the "prescribed voltage" means—

(a) 1000 volts AC or 1500 volts DC if measured between any two conductors; or

(b) 600 volts AC or 900 volts DC if measured between a conductor and earth.

(3) No apparatus to which this article applies is to be installed unless it is provided with a cut-off switch.

(4) Subject to paragraph (5), the cut-off switch must be so placed, and coloured or marked as to satisfy such reasonable requirements as the fire and rescue authority may impose to secure that it must be readily recognisable by and accessible to fire-fighters.

(5) If a cut-off switch complies in position, colour and marking with the current regulations of the Institution of Electrical Engineers for a fire-fighter's emergency switch, the fire and rescue authority may not impose any further requirements pursuant to paragraph (4).

(6) Not less than 42 days before work is begun to install apparatus to which this article applies, the responsible person must give notice to the fire and rescue authority showing where the cut-off switch is to be placed and how it is to be coloured or marked.

(7) Where notice has been given to the fire and rescue authority as required by paragraph (6), the proposed position, colouring or marking of the switch is deemed to satisfy the requirements of the fire authority unless, within 21 days from the date of the service of the notice, the fire and rescue authority has served on the responsible person a counter-notice stating that their requirements are not satisfied.

(8) Where apparatus to which this article applies has been installed in or on premises before the day on which this article comes into force, the responsible person must, not more than 21 days after that day, give notice to the fire and rescue authority stating whether the apparatus is already provided with a cut-off switch and, if so, where the switch is placed and how it is coloured or marked.

(9) Subject to paragraph (10), where apparatus to which this article applies has been installed in or on premises before the day on which this article comes into force, the fire and rescue authority may serve on the responsible person a notice—

(a) in the case of apparatus already provided with a cut-off switch, stating that they are not satisfied with the position, colouring or marking of the switch and requiring the responsible person, within such period as may be specified in the notice, to take such steps as will secure that the switch will be so placed or coloured or marked as to be readily recognisable by, and accessible to, fire-fighters in accordance with the reasonable requirements of the fire and rescue authority; or

(b) in the case of apparatus not already provided with a cut-off switch, requiring him, within such period as may be specified in the notice, to provide such a cut-off switch in such a position and so coloured or marked as to be readily recognisable by, and accessible to, fire-fighters in accordance with the reasonable requirements of the fire and rescue authority.

(10) If a cut-off switch complies in position, colour and marking with the current regulations of the Institution of Electrical Engineers for a fire-fighter's emergency switch, the fire and rescue authority may not serve a notice in respect of it under paragraph (9).

(11) This article does not apply to—

(a) apparatus installed or proposed to be installed in or on premises in respect of which a premises licence under the Licensing Act 2003 has effect authorising the use of premises for the exhibition of a film, within the meaning of paragraph 15 of Schedule 1 to that Act; or

(b) apparatus installed in or on premises before the day on which this article comes into force where, immediately before that date—

(i) the apparatus complied with section 10(2) and (3) (requirement to provide cut-off switch) of the Local Government (Miscellaneous Provisions) Act 1982; and

(ii) the owner or occupier of the premises, as the case may be, had complied with either subsection (5) or subsection (7) (notice of location and type of switch) of section 10 of that Act.

Maintenance of measures provided for protection of fire-fighters

38.—(1) Where necessary in order to safeguard the safety of fire-fighters in the event of a fire, the responsible person must ensure that the premises and any facilities, equipment and devices provided in respect of the premises for the use by or protection of fire-fighters under this Order or under any other enactment, including any enactment repealed or revoked by this Order, are subject to a suitable system of maintenance and are maintained in an efficient state, in efficient working order and in good repair.

(2) Where the premises form part of a building, the responsible person may make arrangements with the occupier of any premises forming part of the building for the purpose of ensuring that the requirements of paragraph (1) are met.

(3) Paragraph (2) applies even if the other premises are not premises to which this Order applies.

(4) The occupier of the other premises must co-operate with the responsible person for the purposes of paragraph (2).

(5) Where the occupier of the other premises is not also the owner of those premises, the reference to the occupier in paragraphs (2) and (4) are to be taken to be references to both the occupier and the owner.

Civil liability for breach of statutory duty

39. —(1) Subject to paragraph (2), nothing in this Order is to be construed as conferring a right of action in any civil proceedings (other than proceedings for recovery of a fine).

(2) Notwithstanding section 86 of the Fires Prevention (Metropolis) Act 1774, breach of a duty imposed on an employer by or under this Order, so far as it causes damage to an employee, confers a right of action on that employee in civil proceedings.

Duty not to charge employees for things done or provided

40. No employer may levy or permit to be levied on any employee of his any charge in respect of anything done or provided in pursuance of any requirement of this Order or of regulations made under it.

Duty to consult employees

41. —(1) In regulation 4A of the Safety Representatives and Safety Committees Regulations 1977 (employer's duty to consult and provide facilities and assistance), in paragraph (1)(b), for "or regulation 4(2)(b) of the Fire Precautions (Workplace) Regulations 1997" substitute "or article 13(3)(b) of the Regulatory Reform (Fire Safety) Order 2005".

(2) In regulation 3 of the Health and Safety (Consultation with Employees) Regulations 1996 (duty of employer to consult), in paragraph (b), for "or regulation 4(2)(b) of the Fire Precautions (Workplace) Regulations 1997" substitute "or article 13(3)(b) of the Regulatory Reform (Fire Safety) Order 2005".

Special provisions in respect of licensed etc. premises

42. —(1) Subject to paragraph (2), where any enactment provides for the licensing of premises in relation to which this Order applies, or the licensing of persons in respect of any such premises—

(a) the licensing authority must ensure that the enforcing authority for the premises has the opportunity to make representations before issuing the licence; and

(b) the enforcing authority must notify the licensing authority of any action that the enforcing authority takes in relation to premises to which the licence relates; but no failure on the part of an enforcing authority to notify under this paragraph shall affect the validity of any such action taken.

(2) Paragraph (1) does not apply where the licensing authority is also the enforcing authority.

(3) In this article and article 43(1)(a)—

(a) "licensing authority" means the authority responsible for issuing the licence; and

(b) "licensing" includes certification and registration and "licence" is to be construed accordingly; and

(c) references to the issue of licences include references to their renewal, transfer or variation.

Suspension of terms and conditions of licences dealing with same matters as this Order

43. —(1) Subject to paragraph (3), paragraph (2) applies if—

(a) an enactment provides for the licensing of premises in relation to which this Order applies, or the licensing of persons in respect of any such premises;

(b) a licence is issued in respect of the premises (whether before or after the coming into force of this Order); and

(c) the licensing authority is required or authorised to impose terms, conditions or restrictions in connection with the issue of the licences.

(2) At any time when this Order applies in relation to the premises, any term, condition or restriction imposed by the licensing authority has no effect in so far as it relates to any matter in relation to which requirements or prohibitions are or could be imposed by or under this Order.

(3) Paragraph (1) does not apply where the licensing authority is also the enforcing authority.

Suspension of byelaws dealing with same matters as this Order

44. Where any enactment provides for the making of byelaws in relation to premises to which this Order applies, then, so long as this Order continues to apply to the premises, any byelaw has no effect in so far as it relates to any matter in relation to which requirements or prohibitions are or could be imposed by or under this Order.

Duty to consult enforcing authority before passing plans

45. —(1) Where it is proposed to erect a building, or to make any extension of or structural alteration to a building and, in connection with the proposals, plans are, in accordance with building regulations, deposited with a local authority, the local authority must, subject to paragraph (3), consult the enforcing authority before passing those plans.

(2) Where it is proposed to change the use to which a building or part of a building is put and, in connection with that proposal, plans are, in accordance with building regulations, deposited with a local authority, the authority must, subject to paragraph (3), consult with the enforcing authority before passing the plans.

(3) The duty to consult imposed by paragraphs (1) and (2)—

(a) only applies in relation to buildings or parts of buildings to which this Order applies, or would apply following the erection, extension, structural alteration or change of use;

(b) does not apply where the local authority is also the enforcing authority.

Other consultation by authorities

46. —(1) Where a government department or other public authority intends to take any action in respect of premises which will or may result in changes to any of the measures required by or under this Order, that department or authority must consult the enforcing authority for the premises before taking that action.

(2) Without prejudice to any power of the court to cancel or modify a notice served by a government department or other authority, no failure on the part of the department or authority to consult under paragraph (1) invalidates the action taken.

(3) In paragraph (1), "public authority" includes an approved inspector within the meaning of section 49 of the Building Act 1984.

Disapplication of the Health and Safety at Work etc. Act 1974 in relation to general fire precautions

47. —(1) Subject to paragraph (2), the Health and Safety at Work etc. Act 1974 and any regulations made under that Act shall not apply to premises to which this Order applies, in so far as that Act or any regulations made under it relate to any matter in relation to which requirements are or could be imposed by or under this Order.

(2) Paragraph (1) does not apply—

(a) where the enforcing authority is also the enforcing authority within the meaning of the Health and Safety at Work etc Act 1974;

(b) in relation to the Control of Major Accident Hazards Regulations 1999.

Service of notices etc.

48. —(1) Any notice required or authorised by or by virtue of this Order to be served on any person may be served on him either by delivering it to him, or by leaving it at his proper address, or by sending it by post to him at that address.

(2) Any such notice may—

(a) in the case of a body corporate, be served on or given to the secretary or clerk of that body; and

(b) in the case of a partnership, be served on or given to a partner or a person having control or management of the partnership business.

(3) For the purposes of this article, and of section 7 of the Interpretation Act 1978 (service of documents by post) in its application to this Order, the proper address of any person is his last known address, except that—

(a) in the case of a body corporate or their secretary or clerk, it is the address of the registered or principal office of that body;

(b) in the case of a partnership or person having control or the management of the partnership business, it is the principal office of the partnership,

and for the purposes of this paragraph the principal office of a company registered outside the United Kingdom or of a partnership carrying on business outside the United Kingdom is their principal office within the United Kingdom.

(4) If the person to be served with or given any such notice has specified an address in the United Kingdom other than his proper address as the one at which he or someone on his behalf will accept notices and other documents, that address is also to be treated for the purposes of this article and section 7 of the Interpretation Act 1978 as his proper address.

(5) Without prejudice to any other provision of this article, any such notice required or authorised to be served on or given to the responsible person in respect of any premises (whether a body corporate or not) may be served or given by sending it by post to him at those premises, or by addressing it by name to the person on or to whom it is to be served or given and delivering it to some responsible individual who is or appears to be resident or employed in the premises.

(6) If the name or the address of the responsible person on whom any such notice is to be served cannot after reasonable inquiry be ascertained by the person seeking to serve it, the document may be served by addressing it to the person on whom it is to be served by the description of "responsible person" for the premises (describing them) to which the notice relates, and by delivering it to some responsible individual resident or appearing to be resident on the premises or, if there is no such person to whom it can be delivered, by affixing it or a copy of it to some conspicuous part of the premises.

(7) Any notice required or authorised to be given to or served on the responsible person or enforcing authority may be transmitted to that person or authority—

(a) by means of an electronic communications network (within the meaning given by section 32 of the Communications Act 2003); or

(b) by other means but in a form that nevertheless requires the use of apparatus by the recipient to render it intelligible.

(8) Where the recipient of the transmission is the responsible person, the transmission has effect as a delivery of the notice to that person only if he has indicated to the enforcing authority on whose behalf the transmission is made his willingness to receive a notice transmitted in the form and manner used.

(9) An indication to an enforcing authority for the purposes of paragraph (8)—

(a) must be given to the authority in any manner it requires;

(b) may be a general indication or one that is limited to notices of a particular description;

(c) must state the address to be used and must be accompanied by any other information which the authority requires for the making of the transmission;

(d) may be modified or withdrawn at any time by a notice given to the authority in any manner it requires.

(10) Where the recipient of the transmission is the enforcing authority, the transmission has effect as a delivery of the notice only if the enforcing authority has indicated its willingness to receive a notice transmitted in the form and manner used.

(11) An indication for the purposes of paragraph (10)—

(a) may be given in any manner the enforcing authority thinks fit;

(b) may be a general indication or one that is limited to notices of a particular description;

(c) must state the address to be used and must be accompanied by any other information which the responsible person requires for the making of the transmission;

(d) may be modified or withdrawn at any time in any manner the enforcing authority thinks fit.

(12) If the making or receipt of the transmission has been recorded in the computer system of the enforcing authority, it must be presumed, unless the contrary is proved, that the transmission—

(a) was made to the person recorded in that system as receiving it;

(b) was made at the time recorded in that system as the time of delivery;

(c) contained the information recorded on that system in respect of it.

(13) For the purposes of this article—

"notice" includes any document or information; and

"transmission" means the transmission referred to in paragraph (7).

Application to the Crown and to the Houses of Parliament

49. —(1) Subject to paragraphs (2) to (4), this Order, except for articles 29, 30 and 32 to 36, binds the Crown.

(2) Articles 27 and 31 only bind the Crown in so far as they apply in relation to premises owned by the Crown but not occupied by it.

(3) For the purposes of this article—

(a) the occupation of any premises by the Corporate Officer of the House of Lords for the purposes of that House, by the Corporate Officer of the House of Commons for the purpose of that House, or by those Corporate Officers acting jointly for the purposes of both Houses, is to be regarded as occupation by the Crown;

(b) any premises in which either or both of those Corporate Officers has or have an interest which is that of an owner are to be regarded as premises owned by the Crown; and

(c) in relation to premises specified in sub-paragraphs (a) and (b), the relevant Corporate Officer is the responsible person.

(4) Nothing in this Order authorises the entry of any premises occupied by the Crown.

(5) Nothing in this Order authorises proceedings to be brought against Her Majesty in her private capacity, and this paragraph shall be construed as if section 38(3) of the Crown Proceedings Act 1947 (interpretation of references in that Act to Her Majesty in her private capacity) were contained in this Order.

Guidance

50. —(1) The Secretary of State must ensure that such guidance, as he considers appropriate, is available to assist responsible persons in the discharge of the duties imposed by articles 8 to 22 and by regulations made under article 24.

(2) In relation to the duty in paragraph (1), the guidance may, from time to time, be revised.

(3) The Secretary of State shall be treated as having discharged his duty under paragraph (1) where—

(a) guidance has been made available before this article comes into force; and

(b) he considers that the guidance is appropriate for the purpose mentioned in paragraph (1).

Application to visiting forces, etc.

51. This Order applies to a visiting force or an international headquarters or defence organisation designated for the purposes of the International Headquarters and Defence Organisations Act 1964 only to the extent that it applies to the Crown.

Subordinate provisions

52. —(1) For the purposes of section 4(3) of the Regulatory Reform Act 2001 (subordinate provisions) the following are designated as subordinate provisions—

(a) article 1(3);

(b) in article 2, the definition of "relevant local authority";

(c) article 9(6) and (7);

(d) in article 10, the reference to "Part 3 of Schedule 1";

(e) article 11(2);

(f) article 14(2);

(g) article 16(1)(a) to (d);

(h) article 16(4);

(i) article 18(6) and (7);

(j) article 25;

(k) article 45(3);

(l) article 49; and

(m) Schedule 1.

(2) A subordinate provisions order made in relation to article 1(3) shall be subject to annulment in pursuance of a resolution of either House of Parliament.

(3) A subordinate provisions order made in relation to any of the provisions mentioned in article 52(1)(b) to (m) may not be made unless a draft of the instrument has been laid before, and approved by a resolution of, each House of Parliament.

Repeals, revocations, amendments and transitional provisions

53. —(1) The enactments and instruments referred to in Schedules 2 and 3 are amended, repealed and revoked in accordance with those Schedules.

(2) The enactments and instruments specified in column 1 of Schedules 4 and 5 are repealed or revoked, as the case may be, to the extent specified in the corresponding entry in column 3.

(3) Any conditions imposed under section 20(2A) or (2C) of the London Building Acts (Amendment) Act 1939 before the date when this Order comes into force and which relate to maintenance, shall cease to have effect from that date.

Appendix B – Fire risk assessment examples

This appendix contains examples of fire risk assessment reports for real buildings, carried out as part of the requirements of the *Fire Precautions (Workplace) Regulations* 1997 (as amended) and the *Management of Health and Safety at Work Regulations* 1999. Whilst still referring to the old legislation and fire certificates, simple replacement of these references with others to the RR(FS)O should suffice to update the report for use with the new system.

The fire risk assessments contained in this Appendix should not be considered as a template for all risk assessments but should at least provide an idea of how the assessment exercise could be documented.

Example 1

The building in question is a five-storey (plus one basement level) office development in London. The building had been issued with a fire certificate under the old system and the company occupying the premises was at the time generally more aware of fire safety than others that the author has encountered. Good management practices and procedures made the risk assessment exercise a simple task and the attention paid by the occupying company to aspects of fire safety are evident by the relatively small number of items requiring attention detailed at the end of the report.

The reader is encouraged to use as much information from the included example as he/she sees fit, but it will always be wise to seek the opinion of the Local Fire and Rescue Services when deciding how much detail to include in the fire risk assessment.

All specific details have been removed from the example report and replaced where necessary with 'XXXXX' to protect the anonymity of the building and the occupying company concerned.

Executive summary

This Fire Risk Assessment report has been commissioned by XXXXXXXXX.

The report has been produced to ensure compliance with the requirements in the *Fire Precautions (Workplace) Regulations* 1997 (as amended) and the *Management of Health and Safety at Work Regulations* which state that all workplaces above a certain occupancy capacity must carry out a fire risk assessment.

Although generally in compliance with applicable standards and legislation, a number of areas require attention:

- Fire doors wedged open and therefore unable to close in the event of a fire.
- Fire action plan signage missing from fifth floor office.
- Fire extinguishers not securely fastened on fifth floor.
- Upper floor fire doors leading to the dedicated means of escape stairs are fitted with key operated locks which can be operated from both sides of the door – no manual bypass mechanism has been provided.
- Fire plan drawings are out of date and do not reflect the current building arrangement/configuration.
- Fire policy and procedure require updating/revision.

All issues requiring attention and their required actions are detailed at the end of Appendix A to this report.

CONTENTS

1 INTRODUCTION

1.1 Terms of reference

The process of fire risk assessment examines various elements within a building in order to assess whether or not additional protective measures are required.

In assessing the risk of fire several areas are considered:

(a) Sources of fuel and ignition (hazard identification)

– The actual hazards present within the building being assessed; without which a fire could not exist (and therefore no hazard).

(b) People and property at risk

– Whether the occurrence of a fire would result in damage, injury or death.

(c) Control measures in place

– An analysis of the current control measures offering mitigation in the event of a fire and their effectiveness.

1.2 Relevant legislation

A fire certificate is required in accordance with the *Fire Precaution Act* 1971.

A recorded fire risk assessment is required under the *Fire Precautions (Workplaces) Regulations* 1997.

1.3 Details of the building

XXXXXXXXXXXXXXXXXXXXXXXXXXXXXXXXXXXXXXXXX

2 OBSERVATIONS

2.1 Hazard identification

2.1.1 Fuel sources

In general fuel sources are assessed as being low.

2.1.1.1 Flammable liquids

No flammable liquids are stored on the premises with the possible exception of cleaning materials all of which will be securely stored in cleaner's cupboards.

2.1.1.2 Flammable gases

No flammable gases are used or stored on the premises.

2.1.1.3 Combustibles

By the nature of the building occupancy there are combustibles present on all floors. The principal combustible inventory consists of paper based materials, reference books, technical drawings, project files and records, etc. all of which are stored in an orderly fashion on shelving or similar.

2.1.2 Ignition sources

Sources of ignition in the premises are assessed as being low.

- The building is deemed a 'no smoking' area.
- There are no 'hot works' carried out on the premises (other than perhaps plant maintenance which would be covered by a permit to work system).
- No oxidising chemicals are used or stored on the premises.
- Electrical devices are employed throughout the premises and upon inspection appear to be in good, safe working order and are regularly checked for electrical safety.

2.2 Persons at risk

(a) Contractors

Contractors may have occasional access to the building in order to perform work or for trade purposes. Access by contractors is controlled by the office management staff.

(b) Visitors

Visitors will have occasional access to all floors. Visitor access is controlled.

(c) Staff

Staff have access to all areas within the building.

(d) Members of the public

Members of the public have no access to any parts of the building other than as visitors – see above.

(e) Disabled

There is a possibility that disabled persons could access the building occasionally as visitors. Facilities for disabled access have been provided throughout the building with the exception of the fifth floor which is inaccessible to wheelchair users.

2.3 Control measures

2.3.1 Means of escape

2.3.1.1 Fire exit doors

The final exit doors leading from the building are in a good state and kept clear of obstructions at all times.

2.3.1.2 Fire exit routes and fire doors

Generally, fire doors are of the self-closing type and provided between office spaces and the circulation and means of escape stairs. There are a small number of doors which are held open under normal circumstances by means of electromagnetic devices. In the event of a fire, the electromagnets are de-energised and the doors are closed by mechanical, automatic closing devices.

During the assessment a number of fire doors were observed to be wedged open by various means. (Basement level by a water fire extinguisher; fifth floor by means of a wooden wedge.)

Whilst it is accepted that this is common practice for ventilation purposes it must be discouraged as the fire safety of the building relies on the fire doors being closed in the event of a fire. With all the best of intentions it is unlikely that the person leaving the office last in the event of an evacuation will be minded to stop and 'un-wedge' the door to allow it to close.

The fire doors leading to the dedicated means of escape stair at fourth and fifth floor levels do not have traditional fire door furniture. They can be locked by means of a key, from either side. There is no manual method of bypassing the lock in the event of an emergency.

Current reliance upon occupants ensuring that the door is unlocked whilst the floor is occupied and locked again by the last person leaving is inadequate. Should a key be lost there is a real possibility that in the event of a fire the door would be rendered useless as a means of escape.

Fire escape routes are generally clear and free from obstructions.

2.3.1.3 Inner rooms

There are a number of small rooms within the main office areas which may be classed as inner rooms in accordance with Approved Document B. These rooms are however provided with extensive glazing and are not permanently occupied; as a consequence the rooms and their use pose no increased risk to the occupants and are deemed acceptable from a fire safety aspect.

2.3.1.4 Emergency lighting

Emergency lighting has been provided throughout the premises and is subject to regular recorded checks and maintenance.

2.3.1.5 Signage

Emergency escape signage is adequately installed throughout the building and is subject to regular checks, and where illuminated, regular maintenance.

Portable fire extinguishers are generally well signed and located in prominent positions throughout the building.

The newly occupied XXXXX floor office was observed to be devoid of a fire plan notice.

2.3.2 Fire detection and alarm

The entire building is covered by an automatic, analogue addressable fire detection and alarm system designed and installed to meet, as a minimum, the requirements of BS5839 Part 1. The life safety systems are designed to promote and accommodate a simultaneous evacuation of all occupants in the event of a fire alarm.

The main fire alarm panel is located at ground floor level in the entrance lobby area adjacent the main entrance doors.

In the event of a fire the following events automatically take place:

- alarm sounders are initiated;
- held open fire doors area released;
- doors locked by solenoids are released.

2.3.3 Portable fire-fighting equipment

A number of CO_2 and water extinguishers are located throughout the building, generally in excess of the requirements of BS5306.

2.3.4 Fire suppression systems

There are no fixed fire suppression systems installed within the building.

2.3.5 Smoke venting

There are no dedicated smoke venting systems within the building although all floors (including the basement) have a high quantity of openable windows.

2.3.6 Pressurisation

None of the areas within the buildings are deliberately pressurised.

2.3.7 Arson

It is not envisaged that any combustible or flammable materials will be accessible to members of the general public. There are no outdoor storage areas associated with the building. Rubbish bins (lidded and wheeled) are located XXXXXXXXXX and are emptied on a regular basis.

2.3.8 Policy and procedure

The building's fire emergency procedures are currently under review and amendment; as such a practical assessment is not possible at this time, but they are expected to fulfil the applicable statutory requirements.

Evacuation and emergency procedures are required to cover all emergency events and provide instruction to all occupants on reaction to the sounding of an alarm; such documentation is currently being reviewed and amended.

2.3.9 Fire management

The Company Health and Safety Policy describes a structured approach to fire safety and the roles and responsibilities of various individuals in so far as fire safety is concerned.

2.3.10 Training

Each Team Secretary is provided with a fire safety dossier which is kept available at all times for other employees to read. The dossier contains procedures and other resources to assist in the event of a fire or evacuation.

Specific training is made available for employees who have a defined responsibility for fire safety and records of all training are kept on file by the building manager.

APPENDIX A – FIRE RISK ASSESSMENT CHECK-SHEETS

Basement level	
Fire Safety Systems within the Area	
Fire warning system (e.g. break glass system, automatic fire detection, etc.)	Covered by building AFD system
Escape lighting (e.g. non-maintained, one hour/three hour duration, hand-held torches, etc.)	Fixed maintained installation
Other (e.g. sprinkler system, gaseous system, etc.)	None

People at risk		
	YES	**NO**
Is the area multi-occupied?		X
Is there varied/shift working?		X
Are there areas where employees/others may be isolated?		X
Do persons sleep in the area?		X

Fire Hazards		
	YES	**NO**
Are there any fire hazards in the area?	X	
If there are, can they be removed/reduced or replaced?	X	X
Specify: IT equipment and electrical distribution room General office paperwork storage		

Structural Features		
	YES	**NO**
Are there any structural features that could promote the spread of fire?		X
If there are, can they be removed/reduced or replaced?		
Specify: n/a		

Horizontal Evacuation	**ADEQUATE?**	
	YES	**NO**
Control measures for any fire hazards within the area	X	
Control and monitoring of the number of occupants	X	
Definition and number of escape routes	X	
Travel distances	X	
Number and widths of exits	X	
Inner room situations	X	
Corridors	X	
Dead-end conditions	X	
Door openings and door fastenings	X	

Sufficient number of stairways	X	
Housekeeping		X
Provision for disabled persons	X	
Vertical Evacuation	**ADEQUATE?**	
	YES	**NO**
Number of stairs sufficient for occupancy	X	
Width of stairs	X	
Width of exits	X	
Stair protection in terms of fire resisting doors and partitions	X	
Door openings and door fastenings	X	
Places of safety from final exits	X	
Housekeeping	X	
Fire Safety Signs and Notices	**ADEQUATE?**	
	YES	**NO**
Exit signs	X	
'Fire door – keep shut' signs	X	
'Fire door – keep locked shut' signs	X	
'Fire exit – keep clear' signs	X	
General fire action notices	X	
Fire-fighting equipment	X	
Door operating signs (e.g. 'Push bar to open')	X	
Fire Warning System	**YES**	**NO**
Will the system alert all the occupants in the event of a fire?	X	
If manual devices such as rotary gongs are provided can the person operating the device do so in a position of safety?	n/a	
Escape Lighting	**YES**	**NO**
Sufficient illumination to see escape routes clearly?	X	
Sufficient illumination to see external escape routes clearly?	X	
Operates on sub-circuit failure?	X	
Illumination at change of level?	X	
Illumination at change of direction?	X	
Illumination to show fire alarm call points and fire-fighting equipment?	X	
Fire-Fighting Equipment	**YES**	**NO**
Is sufficient fire-fighting equipment provided for the area?	X	
Is the fire-fighting equipment appropriate for the risk?	X	
Is the fire-fighting equipment simple in operation?	X	
Has the fire-fighting equipment been checked/inspected by a competent person in the last 12 months?	X	
Does the fire-fighting equipment conform to a recognised standard?	X	
Is the fire-fighting equipment located on escape routes and near exit doors?	X	
Is the fire-fighting equipment securely hung on wall brackets or suitable floor plates?	X	

Ground Floor level		
Fire Safety Systems within the Area		
Fire warning system (e.g. break glass system, automatic fire detection, etc.)	Covered by building AFD system	
Escape lighting (e.g. non-maintained, one hour/three hour duration, hand-held torches, etc.)	Fixed maintained installation	
Other (e.g. sprinkler system, gaseous system, etc.)	None	

People at risk		
	YES	**NO**
Is the area multi-occupied?		X
Is there varied/shift working?		X
Are there areas where employees/others may be isolated?		X
Do persons sleep in the area?		X
Fire Hazards		
	YES	**NO**
Are there any fire hazards in the area?	X	
If there are, can they be removed/reduced or replaced?		X
Specify: IT equipment and electrical distribution room General office paperwork storage		
Structural Features		
	YES	**NO**
Are there any structural features that could promote the spread of fire?	X	
If there are, can they be removed/reduced or replaced?		X
Specify: An opening exists between ground and first floors. This is a feature of the building and both floors are served by the building AFD system		
Horizontal Evacuation	**ADEQUATE?**	
	YES	**NO**
Control measures for any fire hazards within the area	X	
Control and monitoring of the number of occupants	X	
Definition and number of escape routes	X	
Travel distances	X	
Number and widths of exits	X	
Inner room situations	X	
Corridors	X	
Dead-end conditions	X	
Door openings and door fastenings	X	
Sufficient number of stairways	X	

Housekeeping	X	
Provision for disabled persons	X	
Vertical Evacuation	**ADEQUATE?**	
	YES	**NO**
Number of stairs sufficient for occupancy	X	
Width of stairs	X	
Width of exits	X	
Stair protection in terms of fire resisting doors and partitions	X	
Door openings and door fastenings	X	
Places of safety from final exits	X	
Housekeeping	X	
Fire Safety Signs and Notices	**ADEQUATE?**	
	YES	**NO**
Exit signs	X	
'Fire door – keep shut' signs	X	
'Fire door – keep locked shut' signs	X	
'Fire exit – keep clear' signs	X	
General fire action notices	X	
Fire-fighting equipment	X	
Door operating signs (e.g.'Push bar to open')	X	
Fire Warning System	**YES**	**NO**
Will the system alert all the occupants in the event of a fire?	X	
If manual devices such as rotary gongs are provided can the person operating the device do so in a position of safety?		n/a
Escape Lighting	**YES**	**NO**
Sufficient illumination to see escape routes clearly?	X	
Sufficient illumination to see external escape routes clearly?	X	
Operates on sub-circuit failure?	X	
Illumination at change of level?	X	
Illumination at change of direction?	X	
Illumination to show fire alarm call points and fire-fighting equipment?	X	
Fire-Fighting Equipment	**YES**	**NO**
Is sufficient fire-fighting equipment provided for the area?	X	
Is the fire-fighting equipment appropriate for the risk?	X	
Is the fire-fighting equipment simple in operation?	X	
Has the fire-fighting equipment been checked/inspected by a competent person in the last 12 months?	X	
Does the fire-fighting equipment conform to a recognised standard?	X	
Is the fire-fighting equipment located on escape routes and near exit doors?	X	
Is the fire-fighting equipment securely hung on wall brackets or suitable floor plates?	X	

First Floor level	
Fire Safety Systems within the Area	
Fire warning system (e.g. break glass system, automatic fire detection, etc.)	Covered by building AFD system
Escape lighting (e.g. non-maintained, one hour/three hour duration, hand-held torches, etc.)	Fixed maintained installation
Other (e.g. sprinkler system, gaseous system, etc.)	n/a

People at risk		
	YES	**NO**
Is the area multi-occupied?		X
Is there varied/shift working?		X
Are there areas where employees/others may be isolated?		X
Do persons sleep in the area?		X
Fire Hazards		
	YES	**NO**
Are there any fire hazards in the area?	X	
If there are, can they be removed/reduced or replaced?		X
Specify: IT equipment and electrical distribution room General office paperwork storage		
Structural Features		
	YES	**NO**
Are there any structural features that could promote the spread of fire?	X	
If there are, can they be removed/reduced or replaced?		X
Specify: An opening exists between ground and first floors. This is a feature of the building and both floors are served by the building AFD system		

Horizontal Evacuation	ADEQUATE?	
	YES	**NO**
Control measures for any fire hazards within the area	X	
Control and monitoring of the number of occupants	X	
Definition and number of escape routes	X	
Travel distances	X	
Number and widths of exits	X	
Inner room situations	X	
Corridors	X	
Dead-end conditions	X	
Door openings and door fastenings	X	
Sufficient number of stairways	X	

Housekeeping	X	
Provision for disabled persons	X	
Vertical Evacuation	**ADEQUATE?**	
	YES	**NO**
Number of stairs sufficient for occupancy	X	
Width of stairs	X	
Width of exits	X	
Stair protection in terms of fire resisting doors and partitions	X	
Door openings and door fastenings	X	
Places of safety from final exits	X	
Housekeeping	X	
Fire Safety Signs and Notices	**ADEQUATE?**	
	YES	**NO**
Exit signs	X	
'Fire door – keep shut' signs	X	
'Fire door – keep locked shut' signs	X	
'Fire exit – keep clear' signs	X	
General fire action notices	X	
Fire-fighting equipment	X	
Door operating signs (e.g. 'Push bar to open')	X	
Fire Warning System	**YES**	**NO**
Will the system alert all the occupants in the event of a fire?	X	
If manual devices such as rotary gongs are provided can the person operating the device do so in a position of safety?		n/a
Escape Lighting	**YES**	**NO**
Sufficient illumination to see escape routes clearly?	X	
Sufficient illumination to see external escape routes clearly?	X	
Operates on sub-circuit failure?	X	
Illumination at change of level?	X	
Illumination at change of direction?	X	
Illumination to show fire alarm call points and fire-fighting equipment?	X	
Fire-Fighting Equipment	**YES**	**NO**
Is sufficient fire-fighting equipment provided for the area?	X	
Is the fire-fighting equipment appropriate for the risk?	X	
Is the fire-fighting equipment simple in operation?	X	
Has the fire-fighting equipment been checked/inspected by a competent person in the last 12 months?	X	
Does the fire-fighting equipment conform to a recognised standard?	X	
Is the fire-fighting equipment located on escape routes and near exit doors?	X	
Is the fire-fighting equipment securely hung on wall brackets or suitable floor plates?	X	

Second Floor level	
Fire Safety Systems within the Area	
Fire warning system (e.g. break glass system, automatic fire detection, etc.)	Covered by building AFD system
Escape lighting (e.g. non-maintained, one hour/three hour duration, hand-held torches, etc.)	Fixed maintained installation
Other (e.g. sprinkler system, gaseous system, etc.)	n/a

People at risk		
	YES	**NO**
Is the area multi-occupied?		X
Is there varied/shift working?		X
Are there areas where employees/others may be isolated?		X
Do persons sleep in the area?		X

Fire Hazards		
	YES	**NO**
Are there any fire hazards in the area?	X	
If there are, can they be removed/reduced or replaced?		X
Specify: IT equipment and electrical distribution room General office paperwork storage		

Structural Features		
	YES	**NO**
Are there any structural features that could promote the spread of fire?		X
If there are, can they be removed/reduced or replaced?		n/a
Specify: n/a		

Horizontal Evacuation	**ADEQUATE?**	
	YES	**NO**
Control measures for any fire hazards within the area	X	
Control and monitoring of the number of occupants	X	
Definition and number of escape routes	X	
Travel distances	X	
Number and widths of exits	X	
Inner room situations	X	
Corridors	X	
Dead-end conditions	X	
Door openings and door fastenings	X	
Sufficient number of stairways	X	
Housekeeping	X	

Provision for disabled persons	X	
Vertical Evacuation	**ADEQUATE?**	
	YES	**NO**
Number of stairs sufficient for occupancy	X	
Width of stairs	X	
Width of exits	X	
Stair protection in terms of fire resisting doors and partitions	X	
Door openings and door fastenings	X	
Places of safety from final exits	X	
Housekeeping	X	
Fire Safety Signs and Notices	**ADEQUATE?**	
	YES	**NO**
Exit signs	X	
'Fire door – keep shut' signs	X	
'Fire door – keep locked shut' signs	X	
'Fire exit – keep clear' signs	X	
General fire action notices	X	
Fire-fighting equipment	X	
Door operating signs (e.g. 'Push bar to open')	X	
Fire Warning System	**YES**	**NO**
Will the system alert all the occupants in the event of a fire?	X	
If manual devices such as rotary gongs are provided can the person operating the device do so in a position of safety?		n/a
Escape Lighting	**YES**	**NO**
Sufficient illumination to see escape routes clearly?	X	
Sufficient illumination to see external escape routes clearly?	X	
Operates on sub-circuit failure?	X	
Illumination at change of level?	X	
Illumination at change of direction?	X	
Illumination to show fire alarm call points and fire-fighting equipment?	X	
Fire-Fighting Equipment	**YES**	**NO**
Is sufficient fire-fighting equipment provided for the area?	X	
Is the fire-fighting equipment appropriate for the risk?	X	
Is the fire-fighting equipment simple in operation?	X	
Has the fire-fighting equipment been checked/inspected by a competent person in the last 12 months?	X	
Does the fire-fighting equipment conform to a recognised standard?	X	
Is the fire-fighting equipment located on escape routes and near exit doors?	X	
Is the fire-fighting equipment securely hung on wall brackets or suitable floor plates?	X	

Second Floor level	
Fire Safety Systems within the Area	
Fire warning system (e.g. break glass system, automatic fire detection, etc.)	Covered by building AFD system
Escape lighting (e.g. non-maintained, one hour/three hour duration, hand-held torches, etc.)	Fixed maintained installation
Other (e.g. sprinkler system, gaseous system, etc.)	n/a

People at risk		
	YES	**NO**
Is the area multi-occupied?		X
Is there varied/shift working?		X
Are there areas where employees/others may be isolated?		X
Do persons sleep in the area?		X

Fire Hazards		
	YES	**NO**
Are there any fire hazards in the area?	X	
If there are, can they be removed/reduced or replaced?		X
Specify: IT equipment and electrical distribution room General office paperwork storage		

Structural Features		
	YES	**NO**
Are there any structural features that could promote the spread of fire?		X
If there are, can they be removed/reduced or replaced?		n/a
Specify: n/a		

Horizontal Evacuation	ADEQUATE?	
	YES	**NO**
Control measures for any fire hazards within the area	X	
Control and monitoring of the number of occupants	X	
Definition and number of escape routes	X	
Travel distances	X	
Number and widths of exits	X	
Inner room situations	X	
Corridors	X	
Dead-end conditions	X	
Door openings and door fastenings	X	
Sufficient number of stairways	X	
Housekeeping	X	

Provision for disabled persons	X	
Vertical Evacuation	**ADEQUATE?**	
	YES	**NO**
Number of stairs sufficient for occupancy	X	
Width of stairs	X	
Width of exits	X	
Stair protection in terms of fire resisting doors and partitions	X	
Door openings and door fastenings	X	
Places of safety from final exits	X	
Housekeeping	X	
Fire Safety Signs and Notices	**ADEQUATE?**	
	YES	**NO**
Exit signs	X	
'Fire door – keep shut' signs	X	
'Fire door – keep locked shut' signs	X	
'Fire exit – keep clear' signs	X	
General fire action notices	X	
Fire-fighting equipment	X	
Door operating signs (e.g. 'Push bar to open')	X	
Fire Warning System	**YES**	**NO**
Will the system alert all the occupants in the event of a fire?	X	
If manual devices such as rotary gongs are provided can the person operating the device do so in a position of safety?		n/a
Escape Lighting	**YES**	**NO**
Sufficient illumination to see escape routes clearly?	X	
Sufficient illumination to see external escape routes clearly?	X	
Operates on sub-circuit failure?	X	
Illumination at change of level?	X	
Illumination at change of direction?	X	
Illumination to show fire alarm call points and fire-fighting equipment?	X	
Fire-Fighting Equipment	**YES**	**NO**
Is sufficient fire-fighting equipment provided for the area?	X	
Is the fire-fighting equipment appropriate for the risk?	X	
Is the fire-fighting equipment simple in operation?	X	
Has the fire-fighting equipment been checked/inspected by a competent person in the last 12 months?	X	
Does the fire-fighting equipment conform to a recognised standard?	X	
Is the fire-fighting equipment located on escape routes and near exit doors?	X	
Is the fire-fighting equipment securely hung on wall brackets or suitable floor plates?	X	

Third Floor level	
Fire Safety Systems within the Area	
Fire warning system (e.g. break glass system, automatic fire detection, etc.)	Covered by building AFD system
Escape lighting (e.g. non-maintained, one hour/three hour duration, hand-held torches, etc.)	Fixed maintained installation
Other (e.g. sprinkler system, gaseous system, etc.)	n/a

People at risk		
	YES	NO
Is the area multi-occupied?		X
Is there varied/shift working?		X
Are there areas where employees/others may be isolated?		X
Do persons sleep in the area?		X

Fire Hazards		
	YES	NO
Are there any fire hazards in the area?	X	
If there are, can they be removed/reduced or replaced?		X
Specify: IT equipment and electrical distribution room General office paperwork storage		

Structural Features		
	YES	NO
Are there any structural features that could promote the spread of fire?		X
If there are, can they be removed/reduced or replaced?		n/a
Specify: n/a		

Horizontal Evacuation	ADEQUATE?	
	YES	NO
Control measures for any fire hazards within the area	X	
Control and monitoring of the number of occupants	X	
Definition and number of escape routes	X	
Travel distances	X	
Number and widths of exits	X	
Inner room situations	X	
Corridors	X	
Dead-end conditions	X	
Door openings and door fastenings	X	
Sufficient number of stairways	X	
Housekeeping	X	

Provision for disabled persons	X	
Vertical Evacuation	**ADEQUATE?**	
	YES	**NO**
Number of stairs sufficient for occupancy	X	
Width of stairs	X	
Width of exits	X	
Stair protection in terms of fire resisting doors and partitions	X	
Door openings and door fastenings	X	
Places of safety from final exits	X	
Housekeeping	X	
Fire Safety Signs and Notices	**ADEQUATE?**	
	YES	**NO**
Exit signs	X	
'Fire door – keep shut' signs	X	
'Fire door – keep locked shut' signs	X	
'Fire exit – keep clear' signs	X	
General fire action notices	X	
Fire-fighting equipment	X	
Door operating signs (e.g. 'Push bar to open')	X	
Fire Warning System	**YES**	**NO**
Will the system alert all the occupants in the event of a fire?	X	
If manual devices such as rotary gongs are provided can the person operating the device do so in a position of safety?		n/a
Escape Lighting	**YES**	**NO**
Sufficient illumination to see escape routes clearly?	X	
Sufficient illumination to see external escape routes clearly?	X	
Operates on sub-circuit failure?	X	
Illumination at change of level?	X	
Illumination at change of direction?	X	
Illumination to show fire alarm call points and fire-fighting equipment?	X	
Fire-Fighting Equipment	**YES**	**NO**
Is sufficient fire-fighting equipment provided for the area?	X	
Is the fire-fighting equipment appropriate for the risk?	X	
Is the fire-fighting equipment simple in operation?	X	
Has the fire-fighting equipment been checked/inspected by a competent person in the last 12 months?	X	
Does the fire-fighting equipment conform to a recognised standard?	X	
Is the fire-fighting equipment located on escape routes and near exit doors?	X	
Is the fire-fighting equipment securely hung on wall brackets or suitable floor plates?	X	

Fourth Floor level	
Fire Safety Systems within the Area	
Fire warning system (e.g. break glass system, automatic fire detection, etc.)	Covered by building AFD system
Escape lighting (e.g. non-maintained, one hour/three hour duration, hand-held torches, etc.)	Fixed maintained installation
Other (e.g. sprinkler system, gaseous system, etc.)	n/a

People at risk		
	YES	**NO**
Is the area multi-occupied?		X
Is there varied/shift working?		X
Are there areas where employees/others may be isolated?		X
Do persons sleep in the area?		X

Fire Hazards		
	YES	**NO**
Are there any fire hazards in the area?	X	
If there are, can they be removed/reduced or replaced?		X
Specify: IT equipment and electrical distribution room General office paperwork storage		

Structural Features		
	YES	**NO**
Are there any structural features that could promote the spread of fire?		X
If there are, can they be removed/reduced or replaced?		n/a
Specify: n/a		

Horizontal Evacuation	**ADEQUATE?**	
	YES	**NO**
Control measures for any fire hazards within the area	X	
Control and monitoring of the number of occupants	X	
Definition and number of escape routes	X	
Travel distances	X	
Number and widths of exits	X	
Inner room situations	X	
Corridors	X	
Dead-end conditions	X	
Door openings and door fastenings		X
Sufficient number of stairways	X	

Housekeeping	X	
Provision for disabled persons	X	
Vertical Evacuation	**ADEQUATE?**	
	YES	**NO**
Number of stairs sufficient for occupancy	X	
Width of stairs	X	
Width of exits	X	
Stair protection in terms of fire resisting doors and partitions	X	
Door openings and door fastenings	X	
Places of safety from final exits	X	
Housekeeping	X	
Fire Safety Signs and Notices	**ADEQUATE?**	
	YES	**NO**
Exit signs	X	
'Fire door – keep shut' signs	X	
'Fire door – keep locked shut' signs	X	
'Fire exit – keep clear' signs	X	
General fire action notices	X	
Fire-fighting equipment	X	
Door operating signs (e.g. 'Push bar to open')	X	
Fire Warning System	**YES**	**NO**
Will the system alert all the occupants in the event of a fire?	X	
If manual devices such as rotary gongs are provided can the person operating the device do so in a position of safety?		n/a
Escape Lighting	**YES**	**NO**
Sufficient illumination to see escape routes clearly?	X	
Sufficient illumination to see external escape routes clearly?	X	
Operates on sub-circuit failure?	X	
Illumination at change of level?	X	
Illumination at change of direction?	X	
Illumination to show fire alarm call points and fire-fighting equipment?	X	
Fire-Fighting Equipment	**YES**	**NO**
Is sufficient fire-fighting equipment provided for the area?	X	
Is the fire-fighting equipment appropriate for the risk?	X	
Is the fire-fighting equipment simple in operation?	X	
Has the fire-fighting equipment been checked/inspected by a competent person in the last 12 months?	X	
Does the fire-fighting equipment conform to a recognised standard?	X	
Is the fire-fighting equipment located on escape routes and near exit doors?	X	
Is the fire-fighting equipment securely hung on wall brackets or suitable floor plates?	X	

Fifth Floor level	
Fire Safety Systems within the Area	
Fire warning system (e.g. break glass system, automatic fire detection, etc.)	Covered by building AFD system
Escape lighting (e.g. non-maintained, one hour/three hour duration, hand-held torches, etc.)	Fixed maintained installation
Other (e.g. sprinkler system, gaseous system, etc.)	n/a

People at risk		
	YES	**NO**
Is the area multi-occupied?		X
Is there varied/shift working?		X
Are there areas where employees/others may be isolated?		X
Do persons sleep in the area?		X

Fire Hazards		
	YES	**NO**
Are there any fire hazards in the area?	X	
If there are, can they be removed/reduced or replaced?		X
Specify: IT equipment and electrical distribution room General office paperwork storage		

Structural Features		
	YES	**NO**
Are there any structural features that could promote the spread of fire?		X
If there are, can they be removed/reduced or replaced?		n/a
Specify: n/a		

Horizontal Evacuation	**ADEQUATE?**	
	YES	**NO**
Control measures for any fire hazards within the area	X	
Control and monitoring of the number of occupants	X	
Definition and number of escape routes	X	
Travel distances	X	
Number and widths of exits	X	
Inner room situations	X	
Corridors	X	
Dead-end conditions	X	
Door openings and door fastenings	X	
Sufficient number of stairways	X	

Housekeeping	X	
Provision for disabled persons	n/a	
Vertical Evacuation	**ADEQUATE?**	
	YES	**NO**
Number of stairs sufficient for occupancy	X	
Width of stairs	X	
Width of exits	X	
Stair protection in terms of fire resisting doors and partitions	X	
Door openings and door fastenings		X
Places of safety from final exits	X	
Housekeeping	X	
Fire Safety Signs and Notices	**ADEQUATE?**	
	YES	**NO**
Exit signs	X	
'Fire door – keep shut' signs	X	
'Fire door – keep locked shut' signs	X	
'Fire exit – keep clear' signs	X	
General fire action notices		X
Fire-fighting equipment	X	
Door operating signs (e.g. 'Push bar to open')	X	
Fire Warning System	**YES**	**NO**
Will the system alert all the occupants in the event of a fire?	X	
If manual devices such as rotary gongs are provided can the person operating the device do so in a position of safety?		n/a
Escape Lighting	**YES**	**NO**
Sufficient illumination to see escape routes clearly?	X	
Sufficient illumination to see external escape routes clearly?	X	
Operates on sub-circuit failure?	X	
Illumination at change of level?	X	
Illumination at change of direction?	X	
Illumination to show fire alarm call points and fire-fighting equipment?	X	
Fire-Fighting Equipment	**YES**	**NO**
Is sufficient fire-fighting equipment provided for the area?	X	
Is the fire-fighting equipment appropriate for the risk?	X	
Is the fire-fighting equipment simple in operation?	X	
Has the fire-fighting equipment been checked/inspected by a competent person in the last 12 months?	X	
Does the fire-fighting equipment conform to a recognised standard?	X	
Is the fire-fighting equipment located on escape routes and near exit doors?	X	
Is the fire-fighting equipment securely hung on wall brackets or suitable floor plates?		X

Maintenance	ADEQUATE?	
	YES	NO
Fire rated doors, walls and partitions		
Regular checks	X	
Escape routes and exit doors		
Regular checks	X	
Fire safety signs		
Regular checks	X	
Fire warning system	X	
Weekly	X	
Annual	X	
Escape lighting		
Weekly	X	
Monthly	X	
Annual	X	
Fire-fighting equipment		
Monthly	X	
Annual	X	
All by competent person?	X	
Records kept and location of records?	X	

Method for Calling the Fire Service	ADEQUATE?	
	YES	NO
Method for calling the fire service	X	
Specify: Via PSTN after investigation by responsible person (usually the building manager)		

Policy	YES	NO
Covers:		
The action of employees in the event of a fire?	X	
How people will be warned of a fire?	X	
How an evacuation is carried out, including visitors and people with disabilities?	X	
Assembly points?	X	
Procedures for checking the workplace has been evacuated?	X	
Identification of escape routes?	X	
Fire-fighting equipment provided?	X	
Duties and identities of persons with specific responsibilities in the event of a fire?	X	
Where appropriate the isolating of machinery and processes?	n/a	
Specific arrangements for high-risk areas of the workplace?	n/a	
How the Fire Service is called and who is responsible for doing so?	X	
Liaison with the Fire Service on arrival?	X	

Training	YES	NO
Is there a training programme?	X	
Does it cover:		
The action to take on discovering a fire?	X	
How to raise the alarm?	X	
Action to take on hearing the alarm?	X	
Procedures for alerting members of the public and visitors including, where appropriate, directing them to exits?	X	
Arrangements for calling the Fire Service?	X	

Evacuation procedures for everyone in the workplace?	X	
The location, and where appropriate, the use of fire-fighting equipment?	X	
The location of escape routes?	X	
How to open all escape doors?	X	
The importance of keeping fire doors closed?	X	
Where appropriate, how to stop machinery, processes and isolate power supplies in the event of a fire?	n/a	
The reason for not using lifts (except those specifically installed for disabled people) in the event of a fire?	X	
The importance of general fire safety and good housekeeping?	X	

Fire Safety Deficiencies to be Rectified

Fire safety deficiencies to be rectified			
	Deficiency/rectification	**Date to be rectified**	**Date rectified**
1	XXXXX Door leading to base of circulation stair held open by water fire extinguisher. Instruction via memo or email should be circulated to discourage this activity.		
2	XXXXX floor Fire door to XXXXX can be locked from inside or out and would then not be usable in the event of an emergency. Provision of manual lock override or replacement with pushbar system requires investigation.		
3	XXXXX floor General fire action notice missing. Fire notice to be supplied and fixed to wall.		
4	XXXXX floor Extinguishers near means of escape stair not securely fixed to wall or on floor plate. Extinguishers to be hung on brackets or stood on floor plates and signed.		
5	Company fire safety policy requires revision. Amendment of policy is currently ongoing.		
6	Fire plan drawings do not reflect the current building arrangement. Fire plan drawings should be updated to reflect the current building layout, together with the location and details of all fire detection and alarm equipment, fire-fighting provisions, emergency lighting, escape and exit signage, fire doors, etc.		

APPENDIX B – FIRE CERTIFICATE

Attachment

APPENDIX C – FIRE PLANS

Attachment

APPENDIX D – SAFETY POLICY

Attachment

Information used and limitations of the report

This report is based on a survey and existing drawings and information collected during a visit to the site in XXXXXXXXX.

The information contained within this report is for use solely in relation to XXXXXXXXXXXX and should not be used in relation to any other project/location.

This report should be considered a 'live' document and will require periodic updating when usage/ occupancy of the building changes or as required by legislative requirements.

Issue Date Comments

Prepared by: ....

Checked by: .... Approved by: ....

Example 2

The premises in question comprise a single floor of an office block. The building had been issued with a fire certificate under the old system and the company occupying the premises was at the time generally more aware of fire safety than some the author has encountered. Once again, good management practices and procedures made the risk assessment exercise a simple task and the attention paid by the occupying company to aspects of fire safety are evident by the relatively small number of items requiring attention detailed at the end of the report.

The reader will note that this particular example is relatively short and concise due to the size and nature of the premises concerned.

All specific details have been removed from the example report and replaced where necessary with 'XXXXXX' to protect the anonymity of the building and the occupying company concerned.

Executive summary

This Fire Risk Assessment report has been commissioned by XXXXXX Limited for their office situated at XXXXXX.

The report has been produced to demonstrate compliance with the requirements in the *Fire Precautions (Workplace) Regulations* 1997 (as amended) and the *Management of Health and Safety at Work Regulations* which state that all workplaces above a certain occupancy capacity must carry out a fire risk assessment.

Although generally in compliance with applicable standards and legislation, a number of areas require attention.

- Signage to fire escape door.
- Fire extinguishers in general.
- Training for 'responsible' personnel.
- Provision of 'fire packs' to Team Leader, etc. in accordance with Office Policy.

There are a number of areas where information is awaited from the Landlord's representative. It is expected that this information will be made available prior to 'final' issue of this document subsequent to its approval by the Office Management Team.

All issues requiring attention and their required actions are detailed at the end of Appendix A.

CONTENTS

1 INTRODUCTION

1.1 Terms of reference

The process of fire risk assessment examines various elements within a building in order to assess whether or not additional protective measures are required.

In assessing the risk of fire several areas are considered:

(a) Sources of fuel and ignition (hazard identification)

– The actual hazards present within the building being assessed; without which a fire could not exist (and therefore no hazard).

(b) People and property at risk

– Whether the occurrence of a fire would result in damage, injury or death.

(c) Control measures in place

– An analysis of the current control measures offering mitigation in the event of a fire and their effectiveness.

1.2 Relevant legislation

A fire certificate is required in accordance with the *Fire Precaution Act* 1971.
A recorded fire risk assessment is required under the *Fire Precautions (Workplaces) Regulations* 1997.

1.3 Details of the building

XXXXXX is a five-floor property, situated on XXXXXX.

The building is used as office space for a number of individual companies with the Landlord, XXXXXX, responsible for the overall fire safety of the building and, in particular, the common areas.

The building is provided with one circulation stair and one passenger lift for day-to-day use. A dedicated means of escape stair is provided in addition to the general circulation stair.

The building is unsprinklered but is provided with a comprehensive automatic fire detection and alarm system.

2 OBSERVATIONS

2.1 Hazard Identification

2.1.1 Fuel sources

In general fuel sources are assessed as being low.

2.1.1.1 Flammable liquids

No flammable liquids are stored in the office with the possible exception of small quantities of cleaning materials which are stored in the kitchenette.

2.1.1.2 Flammable gases

No flammable gases are used or stored on the premises.

2.1.1.3 Combustibles

By the nature of the office occupancy there are combustibles present. The principal combustible inventory consists of paper-based materials, reference books, technical drawings, project files and records, etc. all of which are stored in an orderly fashion on shelving or similar.

2.1.2 Ignition sources

Sources of ignition in the premises are assessed as being low.

- The building is deemed a 'no smoking' area.
- There are no 'hot works' carried out on the premises (other than perhaps plant maintenance which would be covered by a permit to work system).
- No oxidising chemicals are used or stored on the premises.
- Electrical devices are employed throughout the premises and upon inspection appear to be in good, safe working order and are regularly checked for electrical safety.

2.2 Persons at risk

(a) Contractors

Contractors may have occasional access to the office in order to perform work or for trade purposes. Access by contractors is controlled by the office management staff.

(b) Visitors

Visitors will have occasional access to the office. Visitor access is controlled by the secure access system; the building entrance is protected by magnetically locked doors, operated by 'key-fob' from the outside and a manual button on the inside. There is a CCTV link from the outside of the entrance doors.

(c) Staff

Staff have access to all areas within the office.

(d) Members of the public

Members of the public have no access to any parts of the building other than as visitors – see above.

(e) Disabled

There is a possibility that disabled persons could access the building occasionally as visitors. Facilities for disabled access have been provided throughout the building by the Landlord and the second floor is easily accessible by use of the lift. Means of escape for any disabled occupants will be assisted by one of the nominated 'responsible' persons.

2.3 Control measures

2.3.1 Means of escape

2.3.1.1 Fire exit doors

The final exit doors leading from the building are in a good state and kept clear of obstructions at all times.

2.3.1.2 Fire exit routes and fire doors

Generally, fire doors are of the self-closing type and provided between office spaces and the circulation and means of escape stairs.

Fire escape routes are generally clear and free from obstructions; however a Christmas tree has been temporarily stored in the south stair pending disposal.

2.3.1.3 Inner rooms

There are no rooms within the main office area which would be classed as inner rooms in accordance with Approved Document B other than the server room which, having a small footprint (approximately 2m x 3m) and scarce occupancy, is deemed not to present any additional risk to occupants.

2.3.1.4 Emergency lighting

Emergency lighting has been provided throughout the premises by the Landlord and is subject to regular recorded checks and maintenance.

2.3.1.5 Signage

Emergency escape signage is adequately installed throughout the building and is subject to regular checks, and, where illuminated, regular maintenance.

There is a lack of general fire action signage and the installed fire extinguishers require supplementary signage.

The south escape door requires additional signage as outlined in Appendix A to this report.

2.3.2 Fire detection and alarm

The entire building is covered by an automatic, analogue addressable fire detection and alarm system designed and installed to meet, as a minimum, the requirements of BS5839 Part 1. The life safety systems are designed to promote and accommodate a simultaneous evacuation of all occupants in the event of a fire alarm.

The main fire alarm panel is located at ground floor level in the entrance lobby area adjacent to the main entrance doors.

In the event of a fire, the audible alarm sounders operate automatically throughout the building.

The fire alarm sounders are tested on a weekly basis.

2.3.3 Portable fire-fighting equipment

A number of CO_2 and water extinguishers are located throughout the floor. The extinguishers require either fixing to the wall, or providing with hard stands. Some of the extinguishers are outside their 12-month inspection/test schedule. Refer to Appendix A for further details.

2.3.4 Fire suppression systems

There are no fixed fire suppression systems installed within the premises.

2.3.5 Smoke venting

There are no dedicated smoke venting systems within the office although there is a relatively high quantity of openable windows which will provide ample cross ventilation in the event that smoke clearance is required.

2.3.6 Pressurisation

None of the areas within the building are deliberately pressurised.

2.3.7 Arson

It is not envisaged that any combustible or flammable materials will be accessible to members of the general public. There are no outdoor storage areas associated with the office.

2.3.8 Policy and procedure

The building's fire emergency procedures have recently been drafted and are expected to fulfil the applicable statutory requirements.

Evacuation and emergency procedures are included in the Office Policy Document and provide instruction to all occupants on reaction to the sounding of an alarm.

2.3.9 Fire management

The Company Health and Safety Policy describes a structured approach to fire safety and the roles and responsibilities of various individuals in so far as fire safety is concerned.

2.3.10 Training

Each Team Leader is to be provided with a fire safety dossier which will be kept available at all times for other employees to read. The dossier contains procedures and other resources to assist in the event of a fire or evacuation.

Currently no training has taken place specific to fire safety and the management thereof. The nominated 'responsible' persons (i.e. Fire Wardens) are required to be trained in aspects of fire safety, with refresher courses attended annually or bi-annually as necessary. Records of any training are to be retained for inspection on the premises, ideally in the office 'master' fire dossier.

APPENDIX A – DEFICIENCIES

Fire safety deficiencies to be rectified			
Deficiency/rectification		Date to be rectified	Date rectified
1	**Deficiency:** Christmas tree left in escape stair. **Rectification:** Remove Christmas tree, e-mail occupants that on no account are stairs or escape routes to be used for storage – however temporary.		
2	**Deficiency:** No 'Fire Door – Keep Shut' signs. **Rectification:** Fit appropriate signage.		
3	**Deficiency:** No 'Fire Action Notice' signs. **Rectification:** Procure and fit appropriate signage.		
4	**Deficiency:** Signage only fitted for water-based extinguishers. **Rectification:** Procure and fit appropriate signage for all extinguishers.		
5	**Deficiency:** Not all fire extinguishers over 12 months old have been tested. **Rectification:** Fire extinguishers to be scheduled for maintenance/testing.		
6	**Deficiency:** Fire extinguishers not securely fixed to wall or stood on floor plates. **Rectification:** Provide suitable floor plates or fix extinguishers securely to wall.		
7	**Deficiency:** The Office Fire Safety Policy requires implementation. This includes the compilation of a Fire Dossier, issue of Team Fire Packs, etc. **Rectification:** Agree and implement fully the Policy Document.		
8	**Deficiency:** No training programme exists for 'responsible' persons. **Rectification:** Devise suitable training programme.		

APPENDIX B – OBSERVATION POINTS

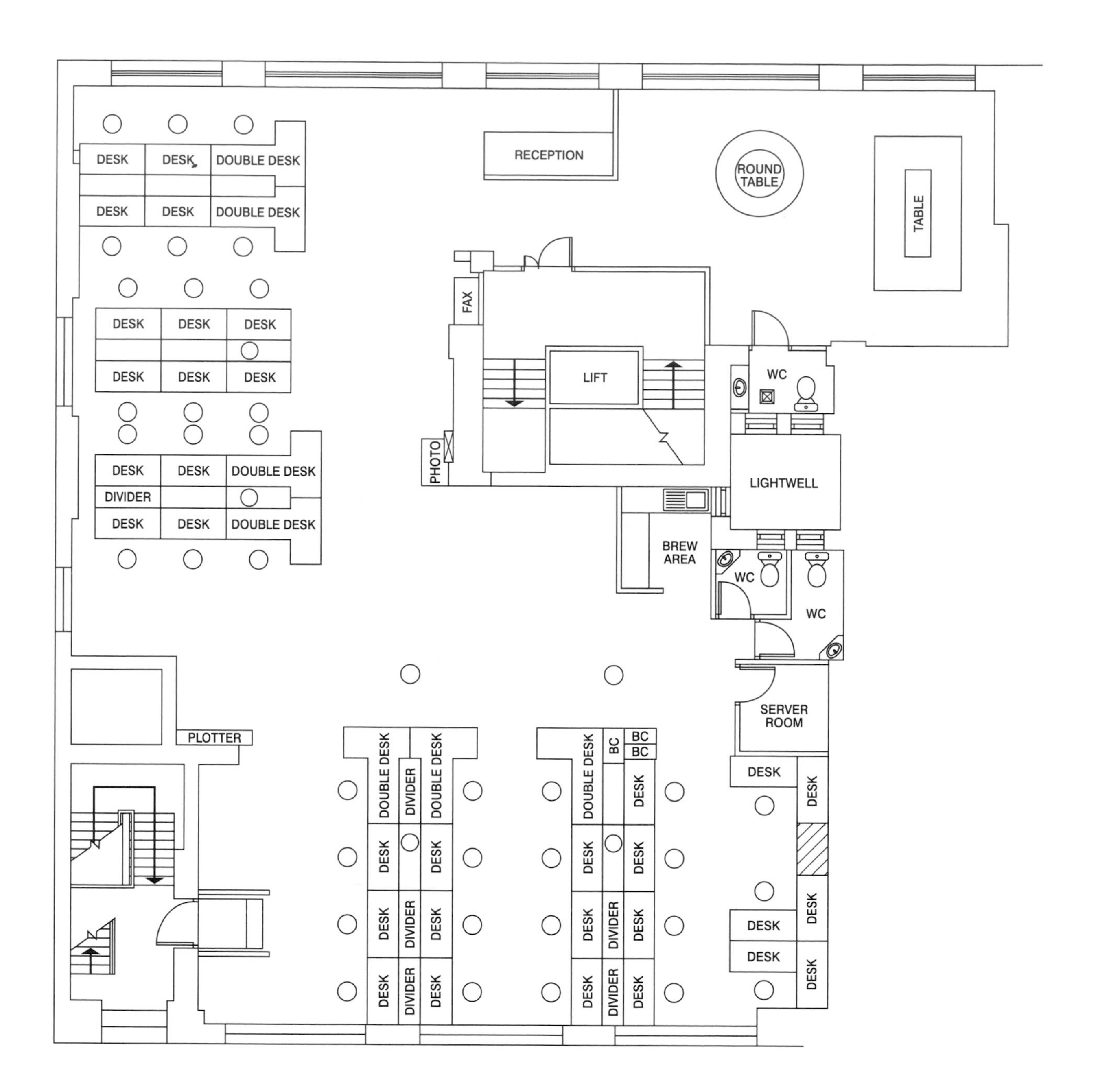

APPENDIX C – FIRE CERTIFICATE

Attachment
(To be supplied by Landlord)

APPENDIX D – FIRE PLAN

APPENDIX E – FIRE SAFETY POLICY

The following text is based on existing XXXXXX policy for the XXXXXX office with minor amendments to permit application to the XXXXXX office.

1 POLICY

It is the policy of XXXXXX to ensure that all employees, visitors and contractors are protected from the risks of fire. In order to achieve this aim, appropriate fire prevention/precaution measures shall be taken; additionally, appropriate evacuation procedures shall be developed, implemented and periodically tested. All persons shall be provided with appropriate fire awareness training and instruction. And all premises shall comply with relevant fire safety legislation and recognised good practice standards.

The main legislation relevant to this subject is the *Fire Precautions Act* 1971 and the *Fire Precautions (Workplace) Regulations* 1997.

2 PROCEDURES/GUIDANCE

2.1 General instruction

All staff must be familiar with the fire procedures as required by the *Fire Precautions Act* 1971, the *Fire Precautions (Workplace) Regulations* 1997 and the *Health and Safety at Work, etc. Act* 1974.
Fire procedures are posted throughout the premises and can be found on exit routes normally adjacent to fire alarm call points or portable fire equipment.

All staff are required to ensure that they are familiar with the alternative means of escape in case of fire by walking the routes from the area in which they are employed.

Occupants should know their assembly point, which is presently located in the car park next to XXXXXX.

If you have to evacuate the premises:

- DO exit quickly and calmly
- DO go directly to open air and proceed to the assembly point
- DO close the door behind you
- DO NOT enter an adjacent building
- DO NOT stop to collect bags
- DO NOT use lifts
- DO NOT attempt to fight a fire unless you have been specifically trained to do so.

Any staff not at their usual work position on hearing the evacuation alarm must leave the building and go to the assembly point. On no account must they return.

'Break glass' fire alarm call points can be found adjacent to storey and final exit doors.

Portable fire extinguishers are sited adjacent to storey and final exits and other strategic locations throughout the premises.

2.2 Fire safety

Fire safety is everyone's responsibility. All employees, visitors and contractors are expected to follow established safety procedures to ensure the safe use of electrical/gas appliances, the safe use, storage and disposal of hazardous/combustible materials and compliance with the requirements of the Company's Smoking Policy.

2.3 Fire precautions

Fire doors must be kept closed at all times to maintain compartmentalisation of the building and to prevent the spread of the fire and/or toxic smoke.

Corridors, stairways, landings and escape routes must be kept clear at all times of anything that is likely to cause a fire or accident or to impede evacuation in an emergency.

Hazardous materials must be stored, used and disposed of in accordance with all legal requirements and safe working practices.

All fire-fighting equipment must be kept free from obstruction, be visibly conspicuous and readily available for use in an emergency. Portable fire-fighting equipment must not be removed or repositioned without authority from the Office Director.

Any obvious or suspected damage to, or misuse of, a fire alarm or fire-fighting equipment must be reported immediately to the Office Director.

2.3.1 Housekeeping

The following general housekeeping rules shall be applied by all staff.

- The photocopier areas on all floors shall be maintained in a safe and tidy condition.
- All paper/documents/drawings shall, as far as practicable, be secured in storage cabinets when not in use and at the end of each working day. To ensure that storage space is not wastefully used, full use of the archiving system shall be made.

2.3.2 Electrical equipment

The Company performs regular electrical safety tests on all Company-owned electrical equipment and any faults found are rectified or the item of equipment replaced.

The Company reserves the right to carry out electrical safety tests on non-Company owned electrical equipment (i.e. desk fans, charger units, etc.) used in the Office.

2.3.3 Fire extinguishers

Fire extinguishers are located in the vicinity of all fire exits and must only be used by Fire Wardens and others trained in their use.

Do not attempt to fight a fire with extinguishers unless you are certain that you can put it out. If you do use an extinguisher make sure that you have a clear exit behind you. Your main priority should be to evacuate the building.

There are two types of extinguishers, both have red bodies but are clearly labelled.

- Water, for use on wood, paper and textile fires (red label).
- Carbon dioxide, for electrical, metal and flammable liquid fires (black label).

2.4 Fire action procedure

Any person suspecting or discovering a fire shall:

- Raise the alarm by breaking the glass of the nearest fire alarm call point
- Follow the procedures below 'on hearing a fire alarm'.

Any person hearing a fire alarm shall:

- Leave the building by the nearest available exit route – DO NOT USE LIFTS
- Close all doors (and windows where safe to do so) behind you as you leave
- Leave the building calmly – DO NOT RUN
- Go directly to your assembly point which is in the car park next to XXXXXX where a roll call will be made by your Team Leader
- Never re-enter the building until instructed to do so by the Office Director or Senior Attending Fire Officer. Never re-enter a building whilst the alarm is still sounding.

Instructions given by the nominated staff (Team Leader/Fire Warden) must be followed and breaches of these procedures will be considered serious and may be dealt with under the Company's Disciplinary Procedures.

2.5 Contacting the Fire Brigade

The Office Director and the Fire Wardens are the nominated persons for establishing contact with the Fire Brigade. In the event of a fire alarm being raised they are responsible for accessing the fire alarm panel in the main entrance lobby, contacting the Fire Brigade and undertaking the following procedures.

1. Dial 999, provide the Operator with the Firm telephone number and request the Fire Brigade.
2. When the Fire Brigade answers, provide the following message distinctly:
   - 'There is a fire at XXXXXX'
   - Inform them of the location of the fire within the building
   - Do not replace the receiver until the address has been repeated by the Fire Brigade.

Thereafter the Office Director and Fire Wardens will be responsible for liaising with the fire wardens and the fire brigade.

2.6 Roles and responsibilities

Within XXXXXX a number of personnel have defined responsibilities relating to fire and fire safety; these are as follows.

2.6.1 Office Director

The Office Director assumes overall responsibility for the generation, administration and implementation of the Company Fire Safety Policy. The Office Director may delegate any or all of his or her responsibility to another person with sufficient and appropriate experience/knowledge.

The Office Director shall:

- be directly responsible for the establishment and maintenance of an effective Fire Safety/Control Plan, take a direct interest in such a Plan and publicly support those to whom he delegates responsibilities for Fire Safety;
- contact the fire brigade in the event of fire or fire alarm being raised;
- provide adequate funds and resources to meet the Fire Safety requirements;
- ensure that all liability is covered by insurance as appropriate, and review insurance and loss records annually;
- ensure that responsibility is properly assigned and accepted at all levels;
- ensure that an annual fire drilling and audit of the Fire Safety/Control Plan is carried out;
- maintain, and periodically update, an effective Fire Safety/Control Plan for the Firm;
- monitor the effective operation of maintenance contracts for Fire Safety Systems and Equipment;
- ensure that manual checks are carried out in accordance with the Fire Certificate requirements and ensure the following checks are carried out:
  - fire extinguishers – monthly inspection to ensure that they are in their proper position, have not been discharged and have suffered no obvious damage;
  - fire alarm – weekly inspection of the panel for normal operation. Weekly operation under alarm conditions;
- appoint and allocate areas for each of the Fire Wardens;
- report building status to Senior Fire Brigade Officer.

2.7 Evacuation procedures for the disabled

2.7.1 Wheelchair users – personal assistant in attendance

On hearing the fire alarm the personal assistant in attendance shall proceed with the wheelchair user to a fire-resisting enclosure, i.e. stairway; the personal assistant once in a fire-resisting stairway should stay with the wheelchair user but must ensure that a message is relayed by the Team Leader, or Fire Warden, to the Office Director who will liaise with the Fire Officer in charge of the first fire appliance to arrive giving the exact location and floor level.

2.7.2 Wheelchair users

The wheelchair user must arrange with their Team Leader that the above procedure is known, and that the Team Leader or Fire Warden has agreed the task of ensuring that the wheelchair user is evacuated or removed safely to a fire-resisting enclosure.

2.7.3 Deaf/hearing impaired persons

There are no visual fire alarm signals within the building; deaf or hearing impaired occupants are encouraged to advise an appropriate member of staff of this fact, so that they may be notified of any alarm.

2.7.4 Blind/visually impaired persons

Blind/visually impaired persons are advised to physically locate evacuation and assembly points in their early days in the Company and should make special arrangements with their Team Leader or a Fire Warden for their evacuation in the event of fire.

2.8 Evacuation drills

In accordance with fire safety legislation, fire evacuation drills will be carried out by the Company at least annually.

The drills will monitor the effectiveness of the local evacuation procedures and, where necessary, identify required changes. They will also time the evacuation and compare the time to a previously determined acceptable time for the particular building, based on national standards and accepted good practice. In cases where the evacuation takes longer than the expected time, a second drill may be carried out at a later date.

Reports on the effectiveness of drills will be produced and presented by the Office Director. Appropriate feedback information will also be published on the Company Intranet.

2.8.1 Fire Wardens

Fire Wardens will be familiar with all the exit points for their area and will direct staff and visitors towards the most appropriate available exit.

It must be stressed that Fire Wardens are not trained to be fire fighters.

In the event that a fire alarm sounds, Fire Wardens are to:

- direct staff and visitors towards the nearest available exit;
- maintain a steady flow of people evacuating the building and prevent 'bottlenecks' building up by redirecting staff and visitors towards other available exits as necessary (so that they are not placed at risk);
- direct staff and visitors away from potential sources of fire, if they are known;
- ensure (so far as is reasonably practicable) that the floor is clear or is actively evacuating;
- leave the building themselves by the nearest available exit;
- report to the Company Secretary on the status of their area (using a predetermined short checklist).

2.8.2 Team Leaders

Team Leaders are responsible for the occupants of their area and must make them aware of the fire procedures for the building.

On hearing the fire alarm the Team Leader must ensure that occupants in his or her area are reminded of the assembly point and that they leave the building by the nearest available exit route in a calm and orderly manner. Team Leaders also undertake the following responsibilities with regard to fires and fire safety:

- maintain an up-to-date Fire Register for their team members;
- keep their Team Fire Pack in a readily accessible location and ensure it contains the following:
  - fire procedure;
  - fire escape plans for XXXXXX;
  - team assembly point notice;
  - Team Fire Register;
  - clip board and pencil.

In the event of an evacuation of the building, each Team Leader will check off people on their Fire Register and advise the Office Director or Fire Warden of anyone unaccountable, including visitors.

2.8.3 Employees

It is the duty of every employee to:

- comply with the building evacuation procedure;
- upon hearing the fire alarm (except for the weekly test), proceed to the assembly point, ensuring their visitors, when present, accompany them;
- comply with instructions from Team Secretaries and Fire Wardens;
- report any violation of fire safety policy to their Team Secretary or a Fire Warden.

2.9 Training, instruction and information

All new employees shall be given local fire safety induction training by their Team Leader (or other appropriate person) in the first month of employment. This will include identification of escape routes, location of fire extinguisher and call points, where the assembly point is and any local hazards that they need to be aware of.

All employees shall be given general fire safety awareness training at least every two years by an appropriate method approved by the Office Director.

The Office Director should ensure that Team Leaders have been identified, and Fire Wardens appointed, and have each been trained to an appropriate level.

The Evacuation Procedures and the Assembly Point location shall be displayed on Fire Action Notices located at strategic points throughout the Building.

2.10 Fire risk assessments

In accordance with fire safety legislation, fire risk assessments shall be organised and or carried out by the Office Director. The risk assessments shall be amended as necessary when circumstances require it (e.g. building changes). The risk assessments shall be reviewed at least annually to ensure their on-going relevance and adequacy.

2.11 Monitoring and audit

Team Leaders and Fire Wardens shall, as part of their day-to-day duties and during inspections, ensure that fire safety precaution and prevention measures are in place and are working as they are intended to. Any deficiencies observed shall be reported to the Office Director for remedial action.

Appendix C – Useful references

The following reference documentation appears in no particular order other than alphabetical by source.

Each of the documents listed has been at some time useful to the author and may or may not be of relevance depending on your particular requirements. Some require a prior knowledge of fire engineering principles, a firm grasp of mathematics and an interest in scientific approaches to solving problems related to fire safety; others are of a more general nature, providing interesting further reading.

The following list is non-exhaustive but represents only an insight into the variety of material available to anyone interested in fire safety and its application to buildings.

BRE
BR 96
Fire safety in buildings

BRE
BR136
Fire and the architect – the communication problem

BRE
BR 186
Design principles for smoke ventilation in enclosed shopping centres

BRE
BR 187
External fire spread – building separation and boundary distances

BRE
BR 225
Aspects of fire precautions in buildings

BRE
BR 258
Design approaches for smoke control in atrium buildings

BRE
BR 367
Fire Modelling

BRE
BR 368
Design methodologies for smoke and heat exhaust ventilation

BRE
BR 375
Natural ventilation in atria for environmental and smoke control – an introduction

BRE
79204
Smoke shafts protecting fire-fighting shafts – their performance and design

BRE
BR 462
Steel structures supporting composite floor slabs in fire

BRE
Fire safety engineering – a reference guide

BSI
BS 5306
Fire extinguishing installations and equipment on premises

BSI
BS5588
Fire precautions in the design, construction and use of buildings

BSI
BS5839
Fire detection and fire alarm systems for buildings

BSI
BS7974
Application of fire engineering principles to the design of buildings

BSI
BS9999
Code of practice for fire safety in the design, construction and use of buildings

BSI
BS476
Fire tests on building materials and structures

BSI
BS EN 12845 – 2003
Fixed fire fighting systems – Automatic sprinkler systems – Design, installation and maintenance

BSI
BS EN 1838 (BS5266 part 7)
Lighting applications – Emergency lighting

BSI
BS EN 81 - 72
Safety rules for the construction and installation of lifts - Particular application for passenger and goods passenger lifts - Part 72 - Fire fighters lifts

BSI
BS EN 3–7
Portable fire extinguishers - Part 7: Characteristics, performance requirements and test methods

CIBSE
Guide E - Fire engineering

DFES
BB 100
Designing and managing against the risk of fire in schools

DCLG
Approved Document B

DCLG
The Building Regulations

FPA
Design guide for the fire protection of buildings - essential principles

FPA
The LPC design guide for the fire protection of buildings

HSE
INDG 163
Five steps to risk assessment

HSE
INDG 218
A guide to risk assessment requirements

HSE
INDG 370
Fire and explosions - how safe is your workplace

HSE
RR 040
Fire risk assessment for workplaces containing flammable substances

NHS
HTM 81
Firecode - Fire precautions in new hospitals

NHS
HTM 83
Firecode - Fire safety in healthcare premises - General fire precautions

NHS
HTM 85
Firecode - Fire precautions in existing hospitals

NHS
HTM 86
Firecode - Fire risk assessment in hospitals

Appendix D – Fire and rescue service contacts

The following list is non-exhaustive but represents website addresses representative of the majority of the UK Fire and Rescue Services.

A

Avon Fire & Rescue Services
Website: www.avonfire.gov.uk

B

Bedfordshire & Luton Fire and Rescue Service
Website: www.bedsfire.com

Buckinghamshire and Milton Keynes Fire & Rescue
Website: www.bucksfire.gov.uk

C

Cambridgeshire Fire & Rescue Service
Website: www.cambsfire.gov.uk

Central Scotland Fire and Rescue Service
Website: www.centralscotlandfire.gov.uk

Cheshire Fire Service
Website: www.cheshirefire.co.uk

Cleveland Fire Brigade
Website: www.clevelandfire.gov.uk

Cornwall County Fire Brigade
Website: www.cornwall.gov.uk

County Durham and Darlington Fire and Rescue Service
Website: www.ddfire.gov.uk

Cumbria Fire and Rescue Service
Website: www.cumbriafire.gov.uk

D

Derbyshire Fire and Rescue Service
Website: www.derbys-fire.gov.uk

Devon Fire and Rescue Service
Website: www.devfire.gov.uk

Dorset Fire & Rescue Service
Website: www.dorsetfire.co.uk

Dumfries & Galloway Fire Brigade
Website: www.dumgal.gov.uk/dumgal/services.aspx

E

East Sussex Fire & Rescue Service
Website: www.esfrs.org

Essex County Fire & Rescue
Website: www.essex-fire.gov.uk

F

Fife Fire & Rescue Service
Website: www.fifefire.gov.uk

G

Gloucestershire Fire and Rescue Service
Website: www.glosfire.gov.uk

Grampian Fire and Rescue Service
Website: www.grampianfirebrigade.co.uk

Greater Manchester Fire and Rescue Service
Website: www.manchesterfire.gov.uk

Guernsey Fire Brigade
Website: www.gov.gg

H

Hampshire Fire and Rescue Service
Website: www.hantsfire.gov.uk

Hereford & Worcester Fire and Rescue Service
Website: www.hwfire.org.uk

Hertfordshire Fire & Rescue
Website: www.hertsdirect.org

Highlands & Islands Fire & Rescue Service
Website: www.hifb.org

Humberside Fire and Rescue Service
Website: www.humbersidefire.gov.uk

I

Isle of Man Fire and Rescue Service
Website: www.iomfire.com

Isle of Wight Fire and Rescue Service
Website: www.iwfire.org.uk

K

Kent Fire & Rescue Service
Website: www.kent.fire-uk.org

L

Lancashire Fire and Rescue Service
Website: www.lancsfirerescue.org.uk

Leicestershire Fire and Rescue Service
Website: www.leicestershire-fire.gov.uk

Lincolnshire Fire and Rescue
Website: www.lincolnshire.gov.uk

London Fire Brigade
Website: www.london-fire.gov.uk

Lothian and Borders Fire and Rescue Service
Website: www.lothian.fire-uk.org

M

Merseyside Fire & Rescue Service
Website: www.merseyfire.gov.uk

Mid and West Wales Fire and Rescue Service
Website: www.mawwfire.gov.uk

N

Norfolk Fire and Rescue Service
Website: www.norfolkfireservice.gov.uk

North Wales Fire and Rescue Authority
Website: www.nwales-fireservice.org.uk

North Yorkshire Fire & Rescue Service
Website: www.northyorksfire.gov.uk

Northamptonshire Fire and Rescue Service
Website: www.northantsfire.org.uk

Northern Ireland Fire & Rescue Service
Website: www.nifrs.org

Northumberland Fire and Rescue Service
Website: www.northumberland.gov.uk

Nottinghamshire Fire & Rescue Service
Website: www.notts-fire.gov.uk

O

Oxfordshire Fire and Rescue Service
Website: www.oxfordshire.gov.uk

R

Royal Berkshire Fire and Rescue Service
Website: www.rbfrs.co.uk

S

Shropshire Fire and Rescue Service
Website: www.shropshirefire.gov.uk

Somerset Fire & Rescue Service
Website: www.somerset.gov.uk

South Wales Fire and Rescue Service
Website: www.southwales-fire.gov.uk

South Yorkshire Fire & Rescue Authority
Website: www.syfire.org.uk

Staffordshire Fire & Rescue
Website: www.staffordshirefire.gov.uk

Strathclyde Fire & Rescue
Website: www.strathclydefire.org

Suffolk Fire and Rescue Service
Website: www.suffolkcc.gov.uk

Surrey Fire and Rescue Service
Website: www.surreycc.gov.uk

T

Tayside Fire & Rescue
Website: www.taysidefire.gov.uk

Tyne & Wear Fire and Rescue Service
Website: www.twfire.gov.uk

W

Warwickshire Fire and Rescue Service
Website: www.warwickshire.gov.uk

West Midlands Fire Service
Website: www.wmfs.net

West Sussex Fire & Rescue Service
Website: www.westsussex.gov.uk

West Yorkshire Fire & Rescue Service
Website: www.westyorksfire.gov.uk

Wiltshire Fire & Rescue Service
Website: www.wfb.org.uk

Index